Planning for an Individual Water System

FOURTH EDITION

Developed by the . . .
AMERICAN ASSOCIATION FOR VOCATIONAL INSTRUCTIONAL MATERIALS
120 Driftmier Engineering Center
Athens, Georgia 30602

. . . in cooperation with
Agricultural Research Service
United States Department of Agriculture
and
Drinking Water Research Division
United States Environmental Protection Agency

The American Association for Vocational Instructional Materials (AAVIM) is a nonprofit national institute.

The institute is a cooperative effort of universities, colleges and divisions of vocational and technical education in the United States and Canada to provide for excellence in instructional materials.

Direction is given by a representative from each of the states, provinces and territories. AAVIM also works closely with teacher organizations, government agencies and industry.

DIRECTOR

W. Harold Parady, Executive Director, AAVIM

REVISION

J. Howard Turner, Editor and Coordinator, AAVIM

GRAPHICS

George W. Smith, Art Director, AAVIM

ACKNOWLEDGMENTS

This publication was issued initially in 1955 under the title *PLANNING WATER SYSTEMS FOR FARM AND HOME.* It was extensively revised and re-issued in 1963. The 1973 edition was completely revised and re-illustrated by AAVIM with the guidance and help of the above mentioned agencies, plus assistance from many colleges, individuals and industrial concerns.

Prepared initially and revised twice by **G. E. Henderson,** formerly Executive Director of AAVIM, in cooperation with **Elmer E. Jones,** Research Agricultural Engineer, Agricultural Research Service, USDA. Illustrations and design are by **George W. Smith, Jr.,** Art Director, AAVIM.

Acknowledgment is given on page 149 to the many individuals, colleges, government agencies, and industries that assisted.

FOURTH EDITION 1982

ISBN 0-89606-097-7

Copyright 1982 by the American Association for Vocational Instructional Materials, 120 Driftmier Engineering Center, Athens, Georgia 30602

Equal Opportunity/Affirmative Action Employer

Printed in the United States of America

Contents

PREFACE 5
INTRODUCTION 7

I. **CHECKING WATER NEEDS AGAINST AVAILABLE SUPPLY** 11
    A. **Determining How Much Water Will Be Needed Daily** 11
    B. **Determining the Adequacy of Existing Water Sources** 13
       Types of Water Sources 13
       Determining the Amount of Water Available 16
       Possibility of Increasing Water Yield 22

II. **DETERMINING WHAT IS NEEDED TO PROVIDE SAFE WATER** 25
    A. **Methods of Protecting Water Sources from Pollution** 26
       Protecting Wells and Springs from Pollution 26
       Reducing Pollution in Cisterns 30
       Reducing Pollution in Ponds and Small Lakes 32
    B. **Determining if Your Water Supply is Safe** 33
    C. **Methods Used for Disinfecting Water** 35
       Disinfecting by the Chlorine Method 35
       Disinfecting by the Pasteurization Method 39
       Disinfecting with the Ultra-violet Light Method ... 41
    D. **What Disinfecting Method to Use** 43
       Effectiveness of Chlorine 43
       Care Required by Disinfecting Units 47
    E. **Providing Protection Against Radioactive Fallout** 49

III. **DETERMINING THE NEED FOR WATER CONDITIONING** 51
     Determining the Probable Cause of Poor Water Quality 52
     Measuring the Extent of the Water-Quality Problem 55
     Methods of Correcting the Water-Quality Problem 55

IV. **UNDERSTANDING WATER-CONDITIONING METHODS** 57
    A. **Means for Controlling Water Hardness** 58
       Reverse-Osmosis Units 58
       Ion-Exchange Units 59
    B. **Means for Controlling Iron in Water** 62
       Phosphate Feeders 62
       Ion-Exchange Units 63
       Oxidizing Filters 63
       Chlorinator-and-filter Units 64
    C. **Means for Controlling Manganese in Water** 64
    D. **Means for Controlling Acid in Water** 64
       Soda-Ash or Caustic-Soda Feeder 64
       Neutralizing Tank 65

E. **Means for Controlling Off-Flavor in Water** 66
   Activated Carbon Filters 66
F. **Means for Controlling Turbidity in Water** 67
   Open Pond Treatment 68
      Coagulation and Sedimentation 68
      Copper-Sulfate Treatment (algae) 68
   Turbidity Treatment Systems 68
      Sedimentation and Filtering System 68
      Diatomite Filter 70
      Rapid Sand Filter 70

V. **DETERMINING WHAT PUMPING INSTALLATION TO MAKE** 73
   A. **What Capacity Pump is Needed** 73
      Capacity Needed to Meet Peak Demands 75
      Capacity Needed for Fire Protection 78
      Rate of Water Yield 79
   B. **Understanding Pump Types** 79
      Principles Involved in Pumping 79
      How a Piston Pump Works 81
      How a Centrifugal Pump Works 83
      How a Centrifugal-Jet (Ejector) Pump Works 85
      How a Turbine Pump Works 86
   C. **What Type of Pump to Use** 87
      Depth to Water 87
      Well Size 87
      Pressure Range Needed for Adequate Water Service 93
      Height Water is Lifted above Pump 94
      Pump Location 95
      Pump Durability and Efficiency 96
      Dealer Service 96
   D. **What Type and Size Water Storage to Use** 97
      Types of Storages and How They Work 97
      Amount of Water to be Stored 100
      Sanitation Features of Different Storages 104
   E. **Understanding Water-System Control Units** ... 104
      Water-Supply Control Switches 105
      Pressure-Tank Air-Volume Controls 107
      Pump Prime Controls 112
      Pressure-Relief Valves 112
      Pressure Gages 113
      Water-Hammer Controls 113
      Sand Removal Units 114
      Safety Check Valves 115
   F. **What Housing to Provide for Pump and Water Storage** 116
      Types of Water-System Installations 116
      Protection from Surface Water (Table 15) 118
      Adequacy of Weather Protection (Table 15) 118
      Adequacy of Drainage 116
      Adequacy of Ventilation 116
      Ease of Cleaning the Enclosure 116
      Ease of Servicing the Equipment 116
      Ease of Servicing the Well 119

VI.	**PLANNING THE PIPING INSTALLATION**		121
	A. Where to Locate Outside Water Outlets		121
	Locating Outlets for Convenience		121
	Locating Outlets for Fire Protection		123
	B. How to Plan the Piping Layout		123
	Determining Water Demand at Various Locations		123
	Determining Where to Place the Pipe Line(s)		125
	C. What Kind of Pipe to Select		128
	Kinds of Piping		128
	Ease of Installation		130
	D. How to Select the Proper Supply-Pipe Size		131
	Determining Pipe Size to Serve One Location		133
	Determining Pipe Size(s) to Serve Two or More Locations		134
	Determining Pipe Sizes for Fire Protection		136
	E. What Pipe Protection to Provide		136
	Protection from Freezing		136
	Protection from Mechanical Damage		138
VII.	**PLANNING THE POWER SOURCE**		139
	A. What Power Source to Use		139
	1. Electrical Power for Pumping Water		139
	2. Windmills for Pumping Water		140
	3. Hydraulic Ram for Pumping Water		140
	4. Other Power Sources		140
	B. What Wiring to Use		141
	Wire Size Needed		141
	Type of Conductor Needed		142
	C. What Electrical Protection to Provide		143
	Feeder Circuit Protection		143
	Pump Motor Protection		144
ACKNOWLEDGMENTS			149
REFERENCES			151
METRIC CONVERSION TABLES			153
INDEX			155

List of Tables

I.	Approximate Daily Water Needs for Home and Farm	12
II.	Net Yield of Water for Cisterns per Square Foot of Catchment Area	21
III.	Capacity of Various Sizes of Cisterns	21
IV.	Guide for Determining Minimum Distance Between Well and Sources of Contamination	30
V.	Comparison of Water Disinfection Methods	44
VI.	Minimum Length and Maximum Size of Pipe Needed to Provide 7 Minutes Contact Time at Full Pump Capacity (Approximate)	46
VII.	Care and Maintenance Required for Disinfection Units	48
VIII.	Analyzing and Measuring Water-Conditioning Problems	53
——.	Maximum Levels of Turbidity, Color and Secondary Contaminant Acceptable for Drinking Water	71
IX.	Pump Sizing Based on Water Demand for Various Uses	77
X.	Pump Selection Chart-Shallow Well, Low Pressure	89
XI.	Pump Selection Chart-Shallow Well, High Pressure	90
XII.	Pumps for Lifting Water from More Than 25 Feet Depth	91-92
XIV.	Pressure Tank Selection Chart	101
XV.	Adequacy of Various Types of Water-System Installations	118
XVI.	Relative Merits of Different Kinds of Piping Materials for Supply Lines	129
XVII.	Copper Circuit-Wire Sizes for Individual Single-Phase Motors	142
XVIII.	Aluminum Circuit-Wire Sizes for Individual Single-Phase Motors	142

Preface

This book tells how to plan for a safe and adequate water supply. It is developed for the person who has the responsibility of providing a water system for a suburban home, vacation home, farm or ranch.

The text is designed especially for the student who may expect to work in the business of sales and service of water systems, including water disinfecting and conditioning equipment. A study of this manual will be helpful to a person engaged in developing water sources such as wells, cisterns and small lakes for domestic water supply.

Anyone who depends on an individual water supply will find this book valuable. It will help him to understand the many factors involved in selecting and using water systems, such as the statutory regulations that may be imposed.

There are no prerequisites to the study of this text. The information is basic and the format is simplified for easy learning.

The general objectives are to help a student become proficient in the following jobs:
1. Determining how much water is needed for persons, livestock, domestic animals, and other uses such as a swimming pool and lawn and garden irrigation.
2. Determining the amount of available water.
3. Determining if the available water is safe.
4. Protecting water sources from pollution.
5. Disinfecting water.
6. Conditioning water.
7. Selecting a pump.
8. Selecting a water storage.
9. Providing housing for the pump and water storage.
10. Selecting the piping system.
11. Selecting the proper wiring.

On completion of your study of this manual, you will be expected to do the following:
1. Tell how much water is needed to meet the needs of a home, as well as outside uses and farm needs, with a specified number of people and livestock.
2. Measure or predict the amount of water available from a well, cistern or small pond.
3. Explain how water becomes polluted and how to prevent this from happening.
4. Describe how to determine if water is safe to use.
5. Describe the methods for disinfecting water, giving the advantages and disadvantages of each method.
6. Tell how to determine the need for water conditioning.
7. Describe the methods of water conditioning and give the advantages and disadvantages of each method.
8. Determine what pumping capacity will be needed to meet peak water-use demands.
9. Determine what type and size pump to select to meet the water-use demands.
10. Make recommendations on how to provide fire protection.
11. Draw a layout of the underground piping system, giving the sizes and dimensions needed to meet the demands.
12. Determine water demand at different locations.
13. Select a power source.
14. Explain what wiring installation is necessary to protect the motor and the motor circuit.

Introduction

FIGURE 1. Few investments provide such varied services as a well-planned water system. With electric service available in almost all rural areas in the nation, almost everyone can now have automatic water service.

Few investments have as little thought given to planning as the installation of a water-supply system for a home or farm. To some, a water system is simply an automatic pump and storage tank that provides water under pressure for the kitchen and bathroom. How much water it supplies, the pressure it maintains, the cleanliness of the water, and provisions for lawn watering, fire protection, etc. are matters which are "discovered" after a crisis has developed.

If the water system is to serve farm needs in addition to home needs, there are profit-making and labor-saving factors to consider—ones such as adequate water for livestock watering, for cleaning floors, for egg production, for cooling in hot weather, and for many other uses.

In this discussion, a **water-supply system** is meant to include the *water source, the pump, the water storage tank,* and the *distribution pipe* to serve the various buildings and hydrants. It also includes *water-treatment equipment,* or *disinfection equipment* that may be needed, as well as the *electrical installation* and provisions for *housing* the equipment.

The first cost may be more for a planned installation than for one that is not planned—you will probably buy a larger pump and a larger tank—but you will save money in the long run, because the different units are properly matched.

If you do not plan your water-supply system, one use is added, then another. Water delivery slows down at one outlet when someone uses water at another. The pump has to be changed for a larger one. A new water source may have to be developed to supply the additional water. Wiring to the motor has to be changed to supply a larger motor. A larger pipe has to be installed. Such are the results of poor planning. It is the most expensive and least satisfactory approach you can use in developing a water-supply system.

You may know how to do plumbing, how to wire, and how to install equipment, but these will not take the place of proper planning.

The value of water under pressure for **home use** is now widely accepted (Figure 1). Little can be said about its advantages for health, safety and convenience that is not now accepted as common knowledge.

The advantages of **water for productive uses** on farms are also widely appreciated, but frequently the actual dollar-and-cents advantages are not known. Here are examples of worth-while productive uses.

Milk production is increased. A summary of studies from several agricultural experiment stations reported in Morrison's "Feeds and Feeding"[1] states:

> Experiments have shown that providing water (for milk cows) by means of automatic drinking bowls, so the cows can drink whenever they wish, increases the yield of good cows 3.5 to 4.0 per cent over watering twice daily, and 6 to 11 per cent over watering once daily.

A study by Jones[2] shows that an adequate water system *can save labor:*

> On one of the dairy farms we studied it was established that an adequate water system could reduce water-related labor requirements 1,000 man hours per year. About two man hours per day could have been saved by use of automatic waterers.

FIGURE 2. No matter how much feed you have available or how good it is, if you limit the drinking water, you are limiting growth and production.

Laying hens are particularly sensitive to lack of water since they have no saliva to help with the swallowing process. That water be available at all times for maximum growth and egg production is reflected in a California report:[3]

> ... when the room temperature was held at 90° F., the following observations were made on laying hens: Withholding water drastically limited feed consumption. The hen conserved her supply of body water by reducing her respiration rate and voiding drier droppings. When water was withheld for 25 hours, only a short interruption of egg production was observed; when water was withheld for 48 hours egg production ceased and some of the pullets molted, but later came back into production; the same was true for the 72-hour period:

Tennessee observations[4] are similar:

> Hens maintained at a temperature of 68° F. and without water for 36 hours dropped from 70% production to 6% production. A mortality of 25% was suffered when hens received no water for a 72-hour period. 93% of the survivors immediately went into a molt.

Several tests have been conducted to determine if egg production can be increased by either *fogging* or *evaporative cooling* (forcing air through moist fabric to secure the cooling effect of evaporation).[5,6,7] There was little or no increase in production, but these methods prevented heat prostration in very hot weather. Chickens start panting when temperatures go above 80° F. and may be subject to a heat stroke when temperatures remain above 100° F. for a period of time.

According to the U.S. Department of Agriculture,[8] *beef production increases* with cool drinking water:

> An increase of 0.4 pound in daily gain of cattle was made by cooling the drinking water from a season-average temperature of 88 degrees at the California Imperial Valley Field Station, El Centro. At times the temperature of drinking water reaches 95 degrees throughout the valley.

Ground water temperatures may vary from as high as 72° F. in the deep south to as low as 37° F. in the northernmost sections of the nation. In most sections of the country, providing cool water is largely a matter of using an automatic waterer supplied by an automatic water system.

Iowa studies[9] show that *limited water limits weight gain with hogs:*

> Hand-watered pigs in the first period (42 days) gained approximately 10 pounds less per pig than the pigs supplied with always-available heated water. Remember that average daily temperatures during this period were low enough to keep water in the unheated troughs frozen most of the time. Thus the slower gains of the hand-watered pigs are attributed to lack of water rather than to any effect of temperature of the water itself.

Protecting hogs from hot weather also increases their gains. In order to offset hot weather:[10]

> Louisiana and Texas experiment stations found that providing a wallow for fattening hogs increased average daily gains of fattening hogs as much as .25 to .40 pound per day over hogs without a wallow.
>
> In a comparison of two kinds of wallows, the Louisiana workers reported no difference in the rate of gain between hogs with sanitary wallow and those provided a fresh-earth wallow. But hogs in the earth wallow became infected with a skin disease which could have caused serious trouble.

More recent studies, conducted in Florida,[11] Georgia,[12] North Carolina,[13] Indiana,[14] and California,[15] generally confirm the Texas and Louisiana studies whether hog wallows or sprinklers are used as the cooling method. In almost all instances, it was noted that gains were made with greater feed efficiency when the hogs were cooled.

A six-year study by Oklahoma[16] showed that watering *increased vegetable production* from 43 percent with eggplant to as much as 238 per cent with snap beans and mango peppers. Other vegetables included in the test were lima beans, 102 per cent; cucumbers, 55 per cent; cantaloupes, 116 per cent; sweet corn, 46 per cent; Irish potatoes, 73 per cent; tomatoes, 49 per cent; and cabbage, 57 per cent. Both overhead and furrow irrigation were used.

From the preceding discussion you can see the importance of having adequate water available when you need it. This is what justifies proper planning. Most water systems in use at the present time do not meet this basic need. They were not adequately planned.

The information given in this publication is presented in the order which should be followed in doing an adequate planning job. Enough details are provided for you to make most of your own decisions. Consequently, when you have completed your study of this information, you should be able to plan a complete water system for yourself or for someone else—a system that will be adequate to meet present needs and foreseeable future needs.

Before a water-supply system is actually installed you will need the help of a good dealer—or experienced specialist—to determine costs, to select the makes of equipment you wish to purchase, and to arrange for installation.

The information you need for planning a safe and adequate water-supply system is provided under the following headings:

    I. Checking water needs against available supply
   II. Determining what is needed to provide safe water
  III. Determining the need for water conditioning
  IV. Understanding water-conditioning methods
   V. Determining what pumping installation to make
  VI. Planning the piping installation
 VII. Planning the power source

FIGURE 3. There are many ways that water can be used productively.

I. Checking Water Needs Against Available Supply

Your first job in planning an adequate water-supply system is to find out *how much water will be needed* for the daily uses you have in mind. With automatic washing and watering devices now available, that figure is almost certain to be higher than any estimate you might make. There have been studies made of water consumption that will help you determine water usage fairly accurately.

Of course, the next question is whether there *is enough water available* to meet your needs. To get the answer to this question may take more time and understanding than is necessary to answer the first one. But, until you get the answer to both questions, planning is impossible.

In this discussion, you will learn how to get the answers to both questions, as they may apply to a wide range of conditions. The information is given under the following headings:

A. Determining how much water will be needed daily

B. Determining the adequacy of existing water sources

A. Determining How Much Water Will Be Needed Daily

Determining your daily water needs is of first importance to proper planning. You need this information to determine how much water your water source—well, spring, cistern or pond—will need to supply to meet your needs.

Several studies have been made of the actual amount of water used in a **home.** It is generally agreed that water usage is increasing. This is partly a result of more liberal general usage, and partly because of the addition of automatic water-using equipment such as dishwashers and clothes washers. This increase has resulted in varying recommendations as to how to figure the amount of water for home usage. The figure for home usage, given in Table I, comes from a Nebraska study.[17]

A more recent national study[18] shows that *lawn sprinkling* has become a large user of water. The amount used varied with the time of the year, the weather, the evaporation rate, the amount of water given off by the plants, and the size of the lawn. On days when watering was done, three to five times more water was used than for normal home use.

The study indicates that *swimming pools,* after they are filled for the season, use about the same amount for maintenance as a lawn area of the same size.

If you have a *garden,* the amount needed is about the same as for lawn sprinkling.[19]

Table I gives the approximate water capacity needed for home uses.

Experiment stations have studied the amount of water needed for watering **livestock, poultry and other uses** over a number of years. These are also shown in Table I.

The figures in Table I give maximum water needs. The reason—water to be of service must be available when you need it. Usually that is in hot weather when gardens and lawns need watering, livestock and poultry are thirsty, and baths are popular.

In looking over the table, you may be surprised at the *large amount of water farm animals consume.* Remember that needs vary with the size of the animal, the water temperature, the air temperature, the amount of water in the feed, the amount of production—in the case of dairy cows and laying hens—and the amount of exercise, particularly in the case of work animals or riding and racing horses.

Another use is **fire protection** which is not figured in your daily needs. It will be considered later, when you select the size pump and water storage you need.

To understand how to use Table I, assume you have a family of four persons. There will be some lawn and garden watering. There is a small swimming pool. There are 30 head of dairy cattle, 40 hogs, and 1,000 laying hens. You would figure needs as follows:

TABLE I. APPROXIMATE DAILY WATER NEEDS FOR HOME AND FARM[19]

	Water Consumption Per Day (gallons)
HOME	
For kitchen and laundry use (including automatic equipment), bathing, sanitary use and other uses inside the home	100 per person
For swimming pool maintenance, per 100 sq. ft.	30
LAWN AND GARDEN	
For lawn sprinkling per 1,000 sq. ft., per sprinkling	600 (approx. 1 in.)
For garden sprinkling, per 1,000 sq. ft., per sprinkling	600 (approx. 1 in.)
FARM (maximum needs)	
Dairy cows (14-15,000 pounds milk) Average drinking rate	20 per head
Dry cows or heifers	15 per head
Calves	7 per head
Beef, yearlings, full feed, 90°F	20 per head
Beef, brood cows	12 per head
Sheep or goats	2 per head
Horses or mules	12 per head
Swine finishing	4 per head
Brood sows, nursing	6 per head
Laying hens (90°F)	9 per 100 birds
Broilers (over 100°F)	6 per 100 birds
Turkeys—15-19 weeks (over 100°F)	20-25 per 100 birds
Ducks*	22 per 100 birds
Dairy sanitation—milk room & milking parlor	500 per day
Flushing floors	10 per 100 sq. ft.
Sanitary hog wallow	100 per day

*Studies on water consumption of ducks were not available. The figure is based on rule-of-thumb method of multiplying amount of feed consumed per day by two. This method is sometimes used for other fowls.

Home

4 persons, 100 gal. per person	400 gal. per day
5,000 sq. ft. of lawn to be sprinkled (5 x 600 gal. per 1,000 sq. ft.)	3,000 gal. per day
1,000 sq. ft. of garden	600 gal. per day
200 sq. ft. of swimming pool (2 x 30)	60 gal. per day

Farm

30 dairy cattle, 20 gal. each	600 gal. per day
Dairy sanitation	500 gal. per day
40 hogs, 4 gal. each	160 gal. per day
1,000 laying hens, 9 gal. per 100 birds	90 gal. per day
Probable maximum water needs—total	5,410 gal. per day

To determine your own total daily water needs, use the table in the same way.

You can see from the example how lawn and garden sprinkling can greatly increase the total water needs for any one day. But, there is a good chance you will do your lawn and garden sprinkling over a two- or three-day period. If over a three-day period, this could lower the water needs per day for lawn and garden sprinkling from 3,000 gallons to 1,000 gallons. This would lower the overall daily needs from 5,410 gallons to 3,410 gallons.

With this information, you can now check your water source(s) to see if there is enough water to meet your maximum daily water needs.

B. Determining the Adequacy of Existing Water Sources

Now that you know the amount of water necessary to meet your daily needs, the next question is whether or not your present water source will *supply the amount of water you need throughout the year.* Your past experience with your well, spring, or other existing source may be enough assurance that it will supply ample water. If so, you are fortunate.

If there is some question about the adequacy of your present source, or if you are developing a new one, you had better check it for adequacy. If it does not supply enough water, *there may be ways of increasing its capacity.* Usually, that is less expensive than developing a new source.

The information and procedures you need in determining the merits of your water source(s) are included under the following headings:
- Types of water sources
- Determining the amount of water available
- Possibility of increasing water yield

Check with your local and state authorities before attempting to dig or drill your own well. Some states require a license for anyone drilling a well.

TYPES OF WATER SOURCES

In some sections of the country there may be a choice of sources which will supply enough water. Other sections may be limited to only one source. The various sources of water include: (a) drilled wells, (b) driven wells, (c) jetted wells, (d) dug wells, (e) bored wells, (f) springs, (g) cisterns, and (h) lakes or ponds. Here is a brief description of each of them.

In some parts of the country wells are practically non existent. State and Federal geological surveys have been made throughout most of the nation to determine the feasibility of finding underground water in a given area. The State and/or Federal Geological Service and local old time well drillers should be contacted before any well development is undertaken. Also, you should check on rural water district possibilities.

A **drilled well** usually consists of a steel pipe (casing), from 4 to 8 inches in diameter extended into the ground until it reaches a satisfactory water source. Much larger wells can be drilled, if necessary. Depths may range from 30 or 40 feet to several hundred feet.

Drilling can be done in soil, gravel, or solid rock. A well-drilling machine is used. There are two types. The *percussion type* punches and hammers its way through soil and rock (Figure 4a). The *rotary type* bites and crushes its way through soil and rock (Figure 4b).

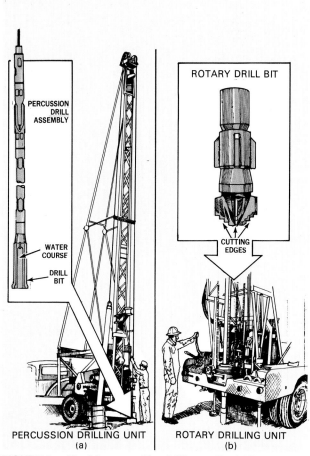

FIGURE 4. Machines used for drilling wells. (a) Percussion unit—one that punches and hammers its way through soil and rock. (Inset) One type of drill used with percussion machines. (b) Rotary unit—one that bites and crushes its way through soil and rock. (Inset) One type of rotary drill bit.

As the hole is drilled, a temporary steel well casing is often used to follow the drill. The temporary casing prevents soil from caving in and filling the hole, if the surrounding formations are not stable. The upper bore hole is made oversize so that cement grout can be added to provide a seal between the bore-hole wall and the permanent casing.

If water at an upper level is unsatisfactory, it is possible to seal it off and drill deeper in search of better quality water or more quantity.

A good drilled well is an excellent source of water because it is usually *free of pollution* and usually provides an *ample supply* of water.

A **driven well** consists of a point and screen attached to a 1¼- to 2-inch pipe and driven into the ground until the screen is below water-table level (Figure 5a). Driven wells are limited to areas where water-bearing sand or gravel lies within about 25 to 100 feet of the surface,

13

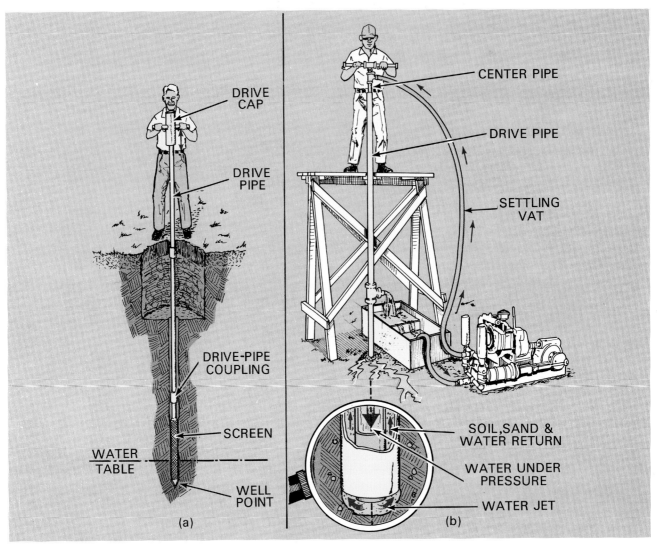

FIGURE 5. (a) A driven well consists of 1¼- to 2-inch pipe, equipped with a point and well screen, and driven below the level of the water table—usually not deeper than 50 feet. (b) A washed or jetted well. These wells may range up to as large as 14 inches in diameter.

and where there are no boulders or rocks to interfere with driving the pipe. This is a comparatively quick and low-cost way to get water if it can be made available by this method.

The *amount of water* any one driven well will supply is usually rather limited. To increase the water supply, it is sometimes possible to connect two or more such wells to one pump.

One type of driven well is called *"washed"* or *"jetted"* well (Figure 5b). This name comes from the way in which the drive pipe is sunk. Instead of the pipe being driven all of the way by mechanical means, a stream of water is forced down a center pipe. The jet action of the water at the lower end of the center pipe dislodges the soil and sand particles at the bottom of the hole. The soil and sand are carried to the surface and discharged by the upward movement of the water passing through the space between the center pipe and the drive pipe. This type of well can be almost any size up to 14 inches in diameter.

A **dug well** consists of a hole in the ground varying from about 3 to 20 feet in diameter, and deep enough to get below the water table. Water will then accumulate in the bottom of the well to about the height of the water table (Figure 6).

Most dug wells are less than 50 feet deep. The sides are cased with tile, stone, or brick masonry.

A **bored well** gets its name from the fact the hole is made with an earth auger (Figure 7). The auger can be hand operated, but power augers are widely used now. The well is usually not more than 8 to 14 inches in diameter, if dug by hand. If power equipment is used, it may be as large as 3 feet in diameter. It is seldom deeper than 100 feet.

A bored well is sunk deeply enough to get below the ground water table. It can be bored to almost any depth as long as boring can be done through sand, gravel, clay, or similar materials, and as long as no rock is encountered. The well is usually lined with vitrified drain tile as the hole is being bored.

FIGURE 6. A dug well consists of a hole 3 to 20 feet in diameter and deep enough to extend below the ground water table.

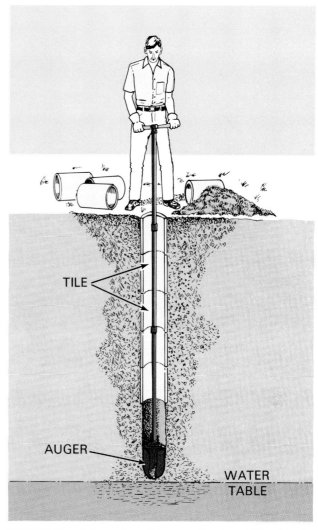

FIGURE 7. A bored well is usually 8 to 14 inches in diameter and extended below the water table.

A **spring** is water that reaches the surface of the ground from some underground supply (Figure 8a). Springs are rather common in rolling or mountainous areas. Some will supply a satisfactory quantity of water, but most of them provide very limited quantities.

A **cistern** usually consists of a watertight, underground reservoir, which is filled with rainwater that drains from the roof of a building, or from several buildings. The cistern provides storage for future use (Figure 8b). Reducing pollution in cisterns is discussed later.

A **lake or pond** consists of water that has accumulated in a depression from surface run-off to form a reservoir. The reservoir may consist of a low place in the surrounding area, or it may be caused by a dam that holds back the water to form a reservoir (Figure 8c). Pond water systems can be expensive to build and maintain. Refer to section "Open Pond Treatment."

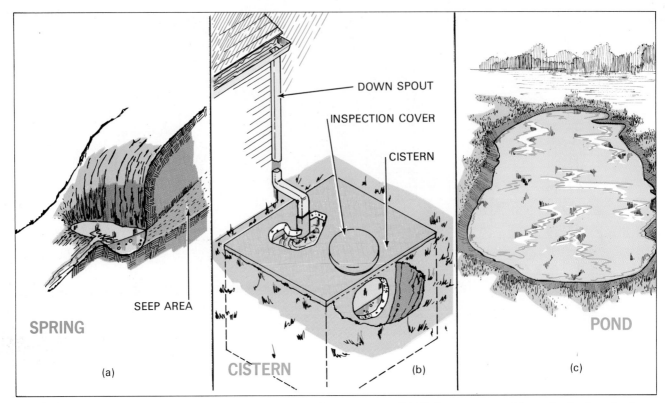

FIGURE 8. Springs, cisterns, and lakes can also be used for supplying water. If you can be assured of a year-round supply, (a) Springs supply water from underground sources. (b) Cisterns are used to collect rainwater from rooftops. (c) Lakes or ponds collect run-off water from surrounding slopes.

DETERMINING THE AMOUNT OF WATER AVAILABLE

After you have selected what appears to be your best water source, it is important to find out about how much water it will provide. The method you use for determining the amount of water a source will provide varies with each type of source.

If you have a **drilled well,** it is probable that the well driller "bailed" the well after it was completed and before a pump was installed. If so, he should have a record of water yield and the amount of drawdown—the amount the water level lowers during pumping.

The *bailer* used by the driller is a long "bucket," or tube, with a valve in the lower end (Figure 9) and a bail on top. To determine well yield, the driller lowers the bailer into the well with his drilling rig, fills it with water, lifts it quickly, and empties the water on the ground. By repeating this over a period of time and counting the number of buckets withdrawn, he can get a good idea as to how much water the well will produce.

He can also determine *drawdown* by marking the cable that holds the bailer when it contacts water on the first draw, then marking the cable again when he finishes. The distance between the two marks on the cable is drawdown, or how much the water level has lowered.

Use of a bailer does not answer the question as to how well the *water* yield will continue during dry periods. That may have to be left to experience. However, drilled wells are ordinarily much deeper than other types of wells, which means they are either less affected by dry conditions or are not affected at all.

If you cannot get water-yield information from your well-driller, and you have no other assurance of yield, you can *test pump* your well. This is the most accurate check you can make. If you do not have a pump of your own, your well driller or local pump dealer can probably supply one. When it is installed, operate it for 4 or 5 hours, if possible—24 hours are better—at a pumping rate somewhat greater than the daily need you determined under "A. Determining How Much Water will be Needed Daily."

If you are not certain about the capacity of the pump you are using for testing, catch water from the pump discharge in buckets or containers of known capacity for a period of 5 minutes. Then multiply that quantity by 12 to get the capacity per hour. Repeat this procedure about every hour—if the water level lowers, well capacity may become less.

If you have a **jetted, driven, bored, or dug well,** your only choice is to use a test pump.

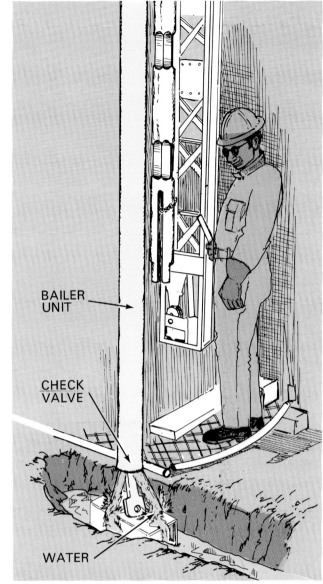

FIGURE 9. A bailer used with a well-drilling rig for determining water yield from a drilled well.

At the same time, be sure to check water-level *drawdown* in the well, both at the start of the pumping operation and as the pumping continues. There are several methods of doing this depending on the type of water source you have.

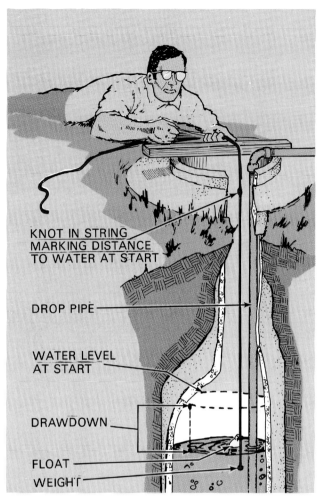

FIGURE 10. String-and-float method for checking drawdown in wells of large diameter.

If there is **room to drop a weighted float** on a string down into the well, drawdown can be checked very simply and easily (Figure 10). Proceed as follows:

1. *Lower weighted float on string until it reaches water level (Figure 10).*
2. *Tie knot in string at surface level before starting the pump.*
3. *Repeat steps 1 and 2 at end of pumping period.*
4. *Measure distance between the upper and lower knots.*

The distance between the two knots represents the drawdown of your well during the pumping period.

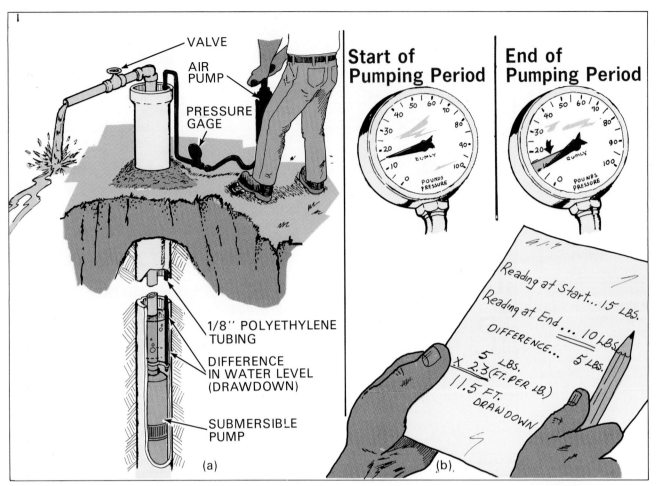

FIGURE 11. (a) Arrangement for checking drawdown in wells of small diameter. (b) Procedure for determining amount of drawdown. (A column of water 2.3 feet high causes a pressure of 1 pound per square inch.)

If your well is too small in diameter to check drawdown with a string and float, you will have to use a water-level indicator (Figure 11). It consists of plastic or copper tubing—⅛-inch diameter—attached to a pressure gage and an air pump. Proceed as follows:

1. *Drop the open end of the tube into the well to a position below the water level (Figure 11a).*

   It is important that the end of the tube not be closer than two feet above the end of the intake to the pump, and it should never be below it. If the end of the tube is too close to the intake, your pressure-gage reading will be wrong because of pump suction.

2. *Pump air into the tube until the pressure-gage reading remains constant.*

   If the pressure-gage pointer returns to zero, there may be a small leak in the tube or at the connections. Check for leak and then replenish the air until the pressure-gage pointer remains constant.

3. *Record pressure-gage reading (Figure 11b).*
4. *Start the pump.*
5. *Check pressure-gage reading at end of pumping period, but while the pump is still running.*
6. *Subtract the lower reading from the higher reading.*

   Most pressure gages are calibrated in pounds per square inch. In Figure 11b, the difference between the two readings is five pounds.

7. *Multiply the reading (in pounds) by 2.3.*

   A column of water that is 2.3 feet high will develop a pressure of one pound. Consequently, by multiplying 2.3 times 5 pounds (example from step 5), you have the total number of feet of drawdown.

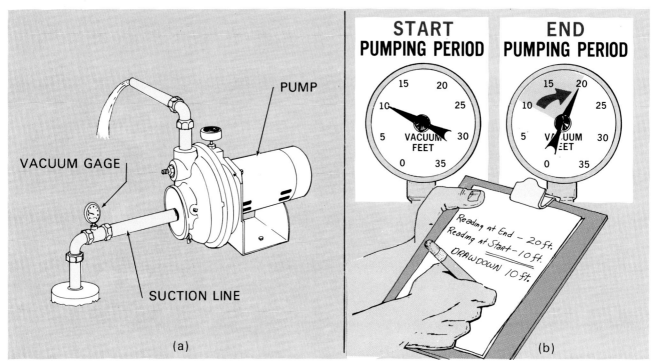

FIGURE 12. (a) Arrangement for checking drawdown on a driven well. (b) Figuring drawdown from vacuum gage readings.

If you have a small-diameter jetted or driven well, the pump suction line may be the only means you can use to check drawdown. In this case you will need to use a **vacuum gage**. Use a piston pump for testing—one that is in good condition. It will lift water by suction as much as 25 feet, under most conditions. Proceed as follows:

1. Connect suction side of pump to well drive pipe (Figure 12a).
2. Insert gage in suction line.
3. Start pumping operation.
4. Check gage reading immediately and record.
5. Check gage reading at end of test period.
6. Subtract lower reading from higher reading (Figure 12b).

Most vacuum gages are *calibrated in feet*. If yours indicates calibration in feet, you have the feet of drawdown already determined.

Some vacuum gages are *calibrated in inches of mercury*. If yours is that type, multiply your reading by 1.13 to get the feet of drawdown.

One precaution—in checking drawdown by this method, it is necessary that all pipe connections be airtight. Otherwise, your reading will not be accurate.

There is no way of telling *how dependable a jetted, driven, bored, or dug well* will be under dry conditions except to depend on your experience, or your neighbor's experience. Since wells of this type are supplied from water that soaks down from the ground surface to the water table, they are usually affected by dry weather.

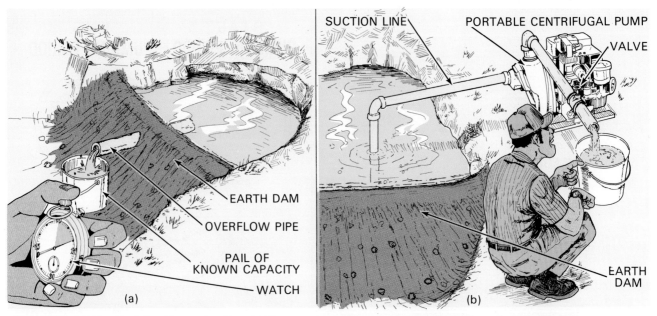

FIGURE 13. Methods of checking spring flow, (a) Measurement from overflow pipe. (b) Use of centrifugal pump.

With **springs,** there is no drawdown problem as there is with wells. But you still need to know (a) how much water it will yield and (b) how well it will keep up its flow in dry weather.

If your spring is on a hillside, the easiest way to check flow is to hollow out a bowl-shaped area at the spring and seal in an overflow pipe near the top with puddled clay. You can then *measure the water yield with buckets* or containers of known size (Figure 13a).

If there is solid rock around the spring, you may have to make a complete bowl of puddled clay, with an overflow pipe in the rim.

If neither of these arrangements is satisfactory, you may be able to arrange with your pump dealer for a *test pump* to check the spring. A centrifugal pump is best for this purpose. Other pumps can be used, but the flow from a centrifugal pump can be more easily throttled with a valve on the discharge side of the pump. You can close the valve until the pump is pumping the exact amount of water the spring yields. The water being pumped can then be measured in buckets or containers of known size (Figure 13b).

If you have noticed that your spring varies in the amount of water it yields at different seasons of the year, it is a poor risk as a water source. It may also be a health hazard. If the flow is constant during all seasons of the year, and is supplying more than your maximum daily needs, it is probably a reliable source.

If you are limited to using a **cistern** as your only source of supply, you may have to find a way of reducing the amount of water you use. This is due to the fact that in most areas it is difficult to collect and store enough rainwater to take care of total home, or farm-and-home, needs. In most cases, some other water source is used along with a cistern.

To determine how much a cistern installation will supply, you will need to get the following information:
— The minimum yearly rainfall recorded for your area.
— The longest period of drought recorded for your area.
— The roof area you have available that can be drained into a cistern (Figure 14).

Rain-collecting surfaces are called "catchment areas" in this discussion. Catchment areas include both roof surfaces and paved ground surfaces. The latter are used occasionally in dry regions.

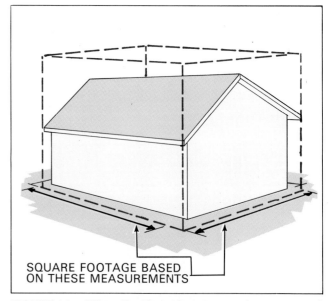

FIGURE 14. When figuring rainwater-catchment area of a roof, measure at ground level below the edges of the roof. Do not measure along the roof slope.

TABLE II. NET YIELD OF WATER FOR CISTERNS PER SQUARE FOOT OF CATCHMENT AREA

Minimum Annual Rainfall	Water Yield (sq. ft.)*
(inches)	(gallons)
10	4.2
15	6.3
20	8.3
25	10.5
30	12.5
35	14.6
40	16.7
45	18.8
50	20.8

*Allows for 1/3 of water being wasted to take care of leakage, roof washing and evaporation. Based on recommendation of Garver, Harry L., "Safe Water for the Farm"; F. B. 1978, 1948.

With this information, you can now get the answer to two questions.
— Is there sufficient rainfall to take care of your yearly needs?
— If so, how large must your cistern be to take care of the worst drought conditions?

To get the answer to these questions:

1. *Multiply your daily needs by 365 (days in year).*
2. *From Table II, select the figure in the column "Minimum Annual Rainfall" that is closest to that of your locality and note the "Water Yield" per square foot you can expect to get.*
3. *Multiply gallons per square foot by the number of square feet of catchment area you have available (Figure 14).*

   This is to determine the amount of water you can depend on collecting each year.

4. *Compare your annual needs (Step 1) with the amount of water that is likely to be available (Step 3).*

   If the total water collected is less than your needs, you will either (a) have to restrict your water use to the amount of water available, or (b) try to develop another water source as a supplement. If the second source is not suitable for bathing and cooking purposes, it could be piped to toilets and outside outlets for sprinkling purposes. If not polluted, it might be used for livestock. A second water pump will be needed.

5. *Determine what size cistern you need.*

   If there is enough water for all needs, multiply your daily water needs by the number of days maximum drought period to get gallons of storage needed.

   If there is enough water for limited use only, multiply your reduced daily needs (estimated) by the number of days maximum drought period to get gallons of storage needed.

6. *Refer to Table III for size cistern(s) needed.*

TABLE III. CAPACITIES OF VARIOUS SIZES OF CISTERNS

Depth in Feet	Diameter of Round Type — Length of Sides of Square Type (Feet)													
	5	6	7	8	9	10	11	12	13	14	15	16	17	18
ROUND TYPE (gallons)														
5	735	1055	1440	1880	2380	2935	3555	4230	4965	5755	6610	7515	8485	9510
6	882	1266	1728	2256	2856	3522	4266	5076	5958	6906	7932	9018	10182	11412
7	1029	1477	2016	2632	3332	4109	4977	5922	6951	8057	9254	10521	11879	13314
8	1176	1688	2304	3008	3808	4696	5688	6768	7944	9208	10576	12024	13576	15216
9	1323	1899	2592	3384	4284	5283	6399	7614	8937	10359	11898	13527	15273	17118
10	1470	2110	2880	3760	4760	5870	7110	8460	9930	11510	13220	15030	16970	19020
12	1764	2532	3456	4512	5712	7044	8532	10152	11916	13812	15864	18036	20364	22824
14	2058	2954	4032	5264	6664	8218	9954	11844	13902	16114	18508	21042	23758	26628
16	2342	3376	4608	6016	7616	9392	11376	13536	15888	18416	21152	24048	27152	30432
18	2646	3798	5184	6768	8568	10566	12798	15228	17874	20718	23796	27054	30546	24236
20	2940	4220	5760	7530	9520	11740	14220	16920	19860	23020	26440	30060	33940	38040
SQUARE TYPE (gallons)														
5	935	1345	1835	2395	3030	3740	4525	5385	6320	7330	8415	9575	10810	12112
6	1122	1614	2202	2874	3636	4488	5430	6462	7584	8796	10098	11490	12974	14534
7	1309	1883	2569	3353	4242	5236	6335	7539	8848	10262	11781	13405	15134	16956
8	1496	2152	2936	3832	4848	5984	7240	8616	10112	11728	13464	15320	17296	19378
9	1683	2421	3303	4311	5454	6732	8145	9693	11376	13194	15147	17235	19458	21800
10	1870	2690	3670	4790	6060	7480	9050	10770	12640	14660	16830	19150	21620	24222
12	2244	3228	4404	5748	7272	8976	10860	12924	15168	17592	20196	22980	25944	29068
14	2618	3766	5138	6706	8484	10472	12670	15078	17696	20524	23562	26810	20268	33912
16	2992	4204	5872	7664	9696	11968	14480	17232	20224	23456	26928	30640	34592	38756
18	3366	4842	6606	8622	10908	13464	16290	19386	22752	26388	30294	34470	38916	42600
20	3740	5380	7340	9580	12120	14960	18100	21540	25280	29320	33660	38300	43240	48444

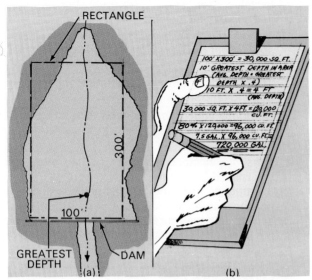

FIGURE 15. Estimating the storage capacity of a small lake. (a) Length and width are established to approximate the size of the lake surface. (b) Steps used to determine the amount of water being stored.

If your primary source of water is a **small lake, pond or reservoir**, which is not recharged frequently by a dependable stream or surface-flowing spring, you will need to *estimate the amount of water* it will supply. Proceed as follows:

1. *Lay out a rectangular shape that is approximately the size of the lake (Figure 15).*
2. *Measure the length and width of the rectangle.*
   In Figure 15a, the example shows a width of 100 feet and a length of 300 feet.
3. *Determine the square feet of lake surface area.*
   Multiply the length by the width. Example (Figure 15):
   100 feet x 300 feet = 30,000 square feet, surface area
4. *Measure the depth of water at the deepest point in the lake.*
   Assume 10 feet for this example.
5. *Determine the average depth of the lake.*
   For small lakes, the average depth is approximately .4 times the greatest depth.
   Example:
   10 feet x .4 = 4 feet average depth (Figure 15b).
6. *Determine the cubic feet of water in the lake.*
   Example:
   30,000 sq. ft. x 4 feet = 120,000 cubic feet
7. *Determine the volume of water available for your use.*
   The volume of water available for your use is usually about 80 per cent of the lake capacity. The remainder is lost through seepage, evaporation and the settling out of sediment.
   Example:
   80% x 120,000 cubic feet = 96,000 cubic feet
8. *Determine number of gallons of water available for your use.*
   Each cubic foot of water is equal to almost 7.5 gallons. Multiply the volume of your lake (cubic feet) by 7.5 gallons.
   Example:
   96,000 cubic feet x 7.5 = 720,000 gallons

You can now check the amount of your daily water needs against the supply available in your lake to see how many days the supply will last. If you have planned to do some yard and garden watering, do not count that as a daily need. Base your needs for those purposes on the number of days you are likely to do that kind of watering.

POSSIBILITY OF INCREASING WATER YIELD

A *well or spring* that has been in use for some time may gradually produce less and less water. There is then a question as to whether the well or spring can be redeveloped to advantage, or is it best to provide a new water source? With *cisterns or ponds* there may be a question of preventing loss through leakage. The approach you make to this problem will depend on the kind of water source you have.

If you have a **drilled well** that has gradually become weaker over a period of time, there is a chance that additional water can be secured by "redevelopment" of the well. This consists of opening up the underground water-supply passages so water may enter the casing with the least amount of difficulty. Water movement into the well casing may be slowed for several reasons. The following are three of the more common ones:

— A buildup of deposits on the well screen or slotted pipe where the water enters the well casing.
— A collection of fine materials — sand, silt and clay — in the water-bearing area around the intake to the well.
— A collection of slime caused by slime-producing bacteria.

Several means have been developed to overcome these problems. The more common methods are overpumping, backwashing, rawhiding, surging, acid/chemical treatment, shock chlorination, and high-velocity jetting. What is involved with each of these is beyond the scope of this discussion. The method used for redevelopment depends on the condition the well driller thinks is causing limited water flow.

Refer to Bulletin 1889 — Water Wells and Pumps — University of California, Berkeley.

Most methods of well redevelopment require special tools, or the use of chemicals, along with considerable experience. Consequently, this discussion is to alert you to the fact that there are several methods for

FIGURE 16. A drilled well equipped with a well screen on the lower end of the casing to provide a large area for the water to enter. Note the coarse sand next to the screen. One method of well development consists of removing the particles, smaller than the well screen openings, so that water can move from the surrounding water-bearing strata with a minimum of resistance.

redeveloping home or farm-size wells. It is not to give instructions on how to redevelop a well yourself.

Another point that must be considered is that some states have *strict laws* regarding well development especially where methods such as dynamiting are used.

Some of the procedures for redeveloping a drilled well can be used for **driven and jetted wells,** but it is usually less costly to drive or jet a new well, or deepen the old one.

Dug and bored wells can sometimes be improved by cleaning out the sediment that accumulates in the bottom of the well after a period of use. They can also be deepened, which may help if the water table in the area is lowering.

If the flow of water from a **spring** is weak, it is sometimes possible to increase it by cleaning the spring so water can pass more easily. Any development beyond that point is highly questionable. Some fairly good springs are ruined by attempts to develop them. Here, again, there is some risk that you may get less water after cleaning. In many areas, the Soil Conservation Service is familiar with ground water conditions and can advise you on the merits of development.

With **cisterns,** the only means to increase the amount of water yield is to increase the cistern capacity or provide more drainage area, or both. If there are indications that water is leaking through the walls of a cistern, it is sometimes possible to apply a heavy plaster film to the inside of the cistern. It acts as a waterproof liner.

Some **ponds** lose considerable water from leakage. With that condition, it is sometimes possible to either add fine silt to the bottom of the pond to help fill and clog the crevices, or use chemical sealant. The latter is mixed with soil and placed on the bottom of the pond to form a water barrier. Get the advice of your local Soil Conservation Service office on this matter.

Notes

II. Determining What is Needed to Provide Safe Water

"Safe water" in this discussion refers to that which is **free from harmful bacteria, viruses, and parasites.** Keeping water "safe" also includes protecting water from radioactive fallout in case of a nuclear explosion. These types of protection are of prime importance to you since they involve the health of your family, and possibly the health of pets or farm animals.

Water-borne-disease-causing-candidates such as giardia cysts may be found in unsafe drinking water.

The U.S. Department of Agriculture in their bulletin, "Safe Water for the Farm," (F. B. No. 1978) states:

Disease-producing bacteria cannot be seen with the naked eye, and thousands may lurk in a drop of water or in a particle of waste matter no larger than a pinhead. Specific germs or parasites in contaminated water may cause typhoid fever, dysentery, diarrhea, or intestinal worms, the more common of which are hookworm, roundworm, whipworm, eelworm, tapeworm and seatworm. The fact that water may be dangerous is seldom realized until after a case of sickness or death.

. . . Contaminated water may contain also the causative agents of ailments common to livestock, such as tuberculosis, hog cholera, anthrax, glanders, and stomach and intestinal worms.

In addition to bacteria and parasites, certain *viruses* can be carried in water such as infectious *hepatitis and polio (poliomyelitis)*.

Of increasing concern is the frequency with which **humans are contracting diseases from animals** through poorly protected water sources (Figure 17). One of the more important of these diseases is leptospirosis, a disease that is common to people and dogs. The disease is often transmitted through dog urine entering the drinking water. Other animal diseases that can be transmitted to humans include anthrax, brucellosis, tuberculosis, and infectious hepatitis.

As the population increases and homes are built closer together in the rural areas, the **problem of pollution becomes greater.** A recent study by the Environmental Protection Agency[20] found that:

Fifty-nine (59) per cent of the rural, individual (water) supplies examined failed to meet the bacteriological standards, and fecal contamination was confirmed in approximately three-fourths of these cases. These systems serve approximately 1,680 people.

The same report also stated:

A sanitary survey of the 571 rural, individual water systems indicates that:

Nearly every system had one or more facility deficiencies.

Very few systems were constructed to prevent entrance of contamination.

Contaminants such as petroleum products, industrial wastes, nitrates from septic tanks, sodium from salted highways and incinerator ash are examples of recent problems with drinking water as reported by the USEPA.

In this discussion *you will learn how to determine if your water supply is polluted and how to protect your water source from pollution.* If you have a lake or pond, there is certain to be some pollution. So you will learn how to keep pollution to the lowest possible level and then how to disinfect the water. You will also learn what precautions are necessary to prevent pollution in case of a nuclear explosion. These points are discussed under the following headings:

A. Methods of Protecting Water Sources from Pollution
B. Determining if Your Water Supply is Safe
C. Methods Used for Disinfecting Water
D. What Disinfecting Method to Use
E. Providing Protection Against Radioactive Fallout

A. Methods of Protecting Water Sources from Pollution

When there is a question about whether a water supply is providing safe drinking water or not, usually the first impulse is to have the health department check a sample. But before a sample is checked, you need to make sure—if it is a **well** or **spring**—that provisions have been made to keep it from being polluted. If not, even though a water sample may check as being satisfactory, the likelihood of future pollution still exists. For that reason the adequacy of well and spring protection must be considered first.

Some wells are much more likely to become polluted than others. For example, water from **dug or bored wells,** which are supplied with water from near the surface, are more likely to become contaminated than those supplied with water from greater depths (Figure 18). But even the best underground water sources can become contaminated if you do not provide the right kind of protection.

If you have a **cistern or pond,** there is certain to be some pollution so it is pointless to test the water. The protective measures given here will help keep down pollution but they will not eliminate it. Generally, the use of cisterns and ponds is discouraged. They should be used only when underground water is not available.

The information given under the following headings will help you understand how to protect wells and springs from pollution and how to keep pollution of other water sources to a minimum:
— Protecting wells and springs from pollution
— Reducing pollution in cisterns
— Reducing pollution in ponds and small lakes

PROTECTING WELLS AND SPRINGS FROM POLLUTION

Pollution of **underground water sources**—wells and springs—is often the result of carelessness. To provide safe water, your first step is to use every means at your disposal to prevent pollution. If you are not successful, then you will need to use some method of water disinfection, which is discussed later.

If an *underground water supply becomes polluted,* you will usually find that one or more of the following conditions exist:
— *The well casing or spring housing is open, or not watertight, at the top. Humans and animals can pollute it directly (Figure 17).*

FIGURE 17. **Wells that are not properly protected at the top can be readily polluted.**

— *The well or spring is located where surface water can drain directly into it (Figure 18).*
— *Water is entering the well after passing through only a few feet of soil (Figure 18). Water that filters through soil is exposed to some pollution at the ground surface. Completeness of filtering will depend on how many feet of soil the water passes through and the character of the soil particles.*
— *The well is being supplied with water from close to the top of the water table (Figure 18).*

The Environmental Protection Agency has this explanation in their bulletin, *Manual of Individual Systems:*[21]

When water seeps downward through overlying material to the water table, particles in suspension, including micro-organisms, may be removed. The extent of removal depends on the thickness and character of the overlying material. Clay or "hardpan" provides the most effective natural protection of ground water. Silt and sand also provide good filtration if fine enough and in thick enough layers. The

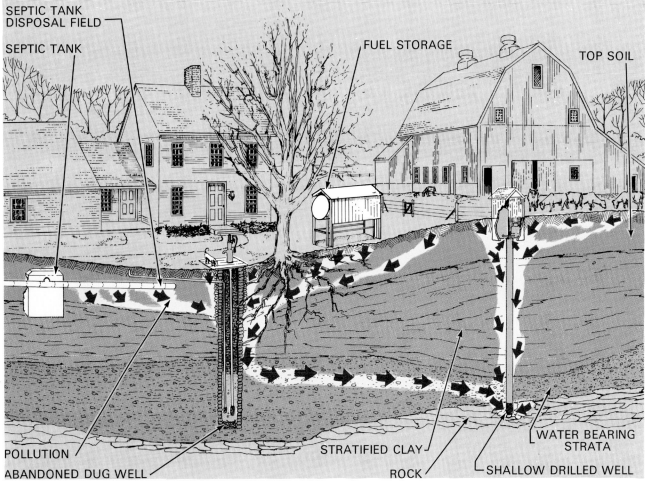

FIGURE 18. Shallow wells can become polluted more readily than deep wells. Note that pollution can come from underground sources as well as from surface sources.

bacterial quality of the water also improves during storage in the aquifer because storage conditions are usually unfavorable for bacterial survival. Clarity alone does not guarantee that ground water is safe to drink; this can only be determined by laboratory testing.

Ground water found in unconsolidated formations (sand, clay, and gravel), and protected by similar materials from sources of pollution, is more likely to be safe than water coming from consolidated formations (limestone, fractured rock, lava, etc.).

— *There is some underground pollution source near the well such as a cesspool, septic tank, privy or an abandoned well (Figure 18).*

Your first concern **in protecting a well from pollution involves three safety precautions.** They are:

— Protection against surface water *entering directly into the top of your well.*

— Protection against surface water *entering below ground level* without filtering through at least 10 feet of earth.

— Protection against surface water *entering the (annular) space* between the well casing and the soil around it.

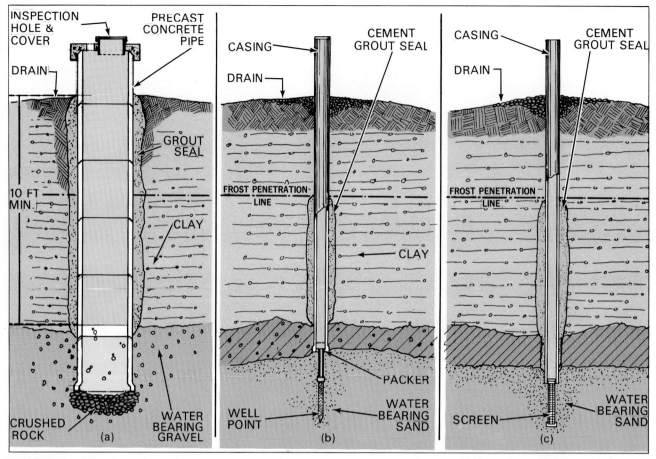

FIGURE 19. Methods of protecting wells from surface pollution, (a) Dug well. (b) Bored well with driven well point. (c) Drilled well.[21]

Figure 19 shows how these different kinds of structural protection can be provided for dug, bored, and drilled wells. Note that protection from surface pollution is accomplished by: (1) extending the casing, or sleeve, 8 to 24 inches above ground level, (2) building up the ground level around the well so it slopes from the well in all directions, and (3) sealing around the well casing with cement grout to keep surface and ground water from moving down through the space between the well casing and the surrounding soil and entering the water-bearing area with very little filtering action.

A very important additional protection is the use of a **sanitary well seal** where the pump connections enter the well. Figure 20 shows two of the more common types.

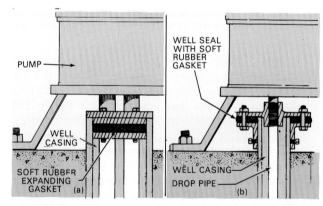

FIGURE 20. Two types of well seals. The seals are designed to keep water from entering the well casing at the top where the pump connections enter. As the steel plates are tightened, the rubber gasket expands against the pipe(s) and well casing to form a seal. (Many states require a pitless adapter instead of a well seal).

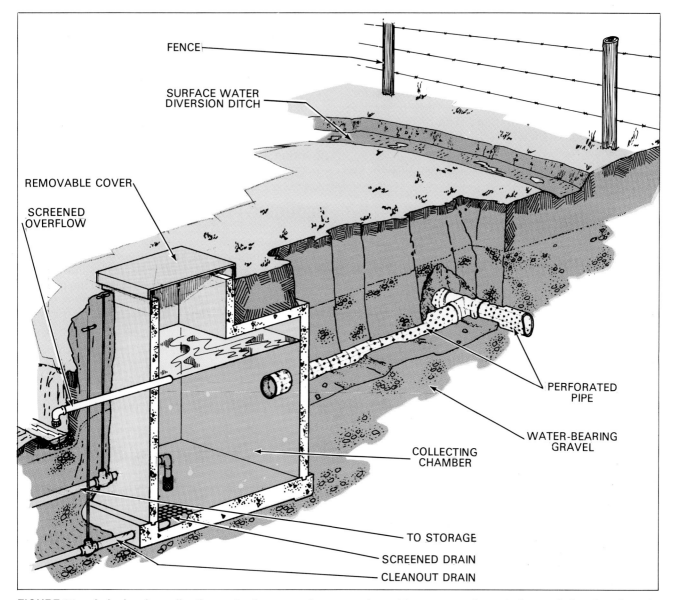

FIGURE 21. A design for collecting water from a spring that provides some protection against surface pollution and provides a means for entering and cleaning the collecting chamber.

In addition to these precautions, if your **well is on the side of a hill,** build a diversion ditch about 50 feet above your pump similar to the one shown for a spring (Figure 21). This keeps surface water from reaching your pump area during heavy rains. Be sure the diversion ditch has enough capacity to take care of run-off from your heaviest rains.

If the **area around your well is open to livestock,** build a fence around it so they cannot get closer than 100 feet.

If you are considering a **new well,** it is important to locate it where it will be as free of pollution as possible. The Environmental Protection Agency recommends the following:[21]

> Since safety of a ground-water source depends primarily on considerations of good well construction and geology, these factors should be the guides in determining safe distances for different situations. The following criteria apply only to properly constructed wells as described in this manual. There is no safe distance for a poorly constructed well.

When a properly constructed well penetrates an unconsolidated formation (sand, clay and gravel) with good filtering properties, and when the aquifer itself is separated from sources of contamination by similar materials, research and experience have demonstrated that 50 feet is an adequate distance separating the two. Lesser distances should be accepted only after a comprehensive sanitary survey, conducted by qualified state or local health agency officials, has satisfied the officials that such lesser distances are both necessary and safe.

If it is proposed to install a properly constructed well in formations of unknown character, the state or U.S. Geological Survey and the state or local health agency should be consulted.

When wells must be constructed in consolidated formations (limestone, fractured rock, lava, etc.), extra care should always be taken in the location of the well and in setting safe distances, since pollutants have been known to travel great distances in such formations. The owner should request assistance from the state or local health agency.

The following table is offered as a guide in determining distances.

TABLE IV. GUIDE FOR DETERMINING MINIMUM DISTANCES BETWEEN WELL AND SOURCES OF CONTAMINATION

Formations	Minimum Acceptable Distance from Well to Source of Contamination
Favorable (unconsolidated)	50 feet Lesser distances only on health department approval following comprehensive sanitary survey of proposed site and immediate surroundings.
Unknown	50 feet only after comprehensive geological survey of the site and its surroundings has established, to the satisfaction of the health agency, that favorable formations do exist.
Poor (consolidated)	Safe distances can be established only following both the comprehensive geological and comprehensive sanitary surveys. These surveys also permit determining the direction in which a well may be located with respect to sources of contamination. In no case should the acceptable distance be less than 50 feet.

After your well is completed, it is almost certain to be contaminated from equipment and materials used to construct it and from surface water that may have entered. It is important that it be *disinfected before you use water from it*. Procedures for disinfecting different kinds of wells are given in the publication "Manual of Individual Water Supply Systems" issued by the U.S. Environmental Protection Agency. It is available from your local health department, or from the address given for reference 21, page 151.

If you have a **spring,** Figure 21 shows a suggested design for collecting the water, storing it, keeping it clean, and providing for cleaning when necessary. It is important that a *removable cover* be provided on top of the spring housing so the collecting compartment may be entered and cleaned occasionally. At the same time, the lid should be heavy enough to discourage anyone from lifting it and dipping a utensil directly into the water.

A *diversion ditch* is important to be sure surface water will not flow into the collecting compartment. In addition, the spring area should be *fenced to keep out livestock* and help avoid possible pollution. Note in Figure 21 that the fence above the spring is located above the diversion ditch and extends around the spring.

NOTE: This system at best is very susceptible to pollution.

REDUCING POLLUTION IN CISTERNS

If your water supply is from a cistern, there is almost **certain to be some pollution** most of the time if not all of the time. The best you can do is to keep the pollution level as low as possible and to use a disinfecting unit to provide safe drinking water. Inspect your cistern periodically.

Pollution can be lowered if the **first rainwater from your roof is wasted.** This washes off most of the bird droppings, dust and accumulated dirt. Wasting the first rainfall is accomplished in two ways: (a) with a hand-operated or (b) an automatic roof wash.

A **diversion valve** (Figure 22a) is normally left in the waste position until enough rain has fallen to wash the roof. The valve is then changed by hand and the water diverted to the cistern. It is not fully satisfactory because of the human tendency to forget to change the valve position as needed.

The **automatic-roofwash** arrangement (Figure 22b) provides for the first rainfall to enter the drum. This collects most of the highly polluted water. After filling the drum, the remaining rainwater automatically enters the cistern. The small opening in the end of the waste pipe on the drum is an **automatic drain.** It is set to drip water until the drum is empty. During a period of no rain, it completely drains the drum, so it is ready to receive the first waste water from the next rain. The sand in the bottom of the drum filters out dirt particles so the small drain hole does not clog. The drum capacity is usually figured at about 10 gallons per 1,000 square feet of catchment area.

Wasting the first rainfall is considered more satisfactory than the use of a sand filter (Figure 23). A **sand filter** can be satisfactory if kept clean, and the filter material changed regularly. However, people do not usually take that kind of care of a filter—the result is more pollution, rather than less. This has caused many state health departments to recommend against their use. Screened filters are also available.

Also, from a user's standpoint, when the filter clogs or when there is heavy rainfall, or both, water is wasted. This is because the supply of incoming water is greater

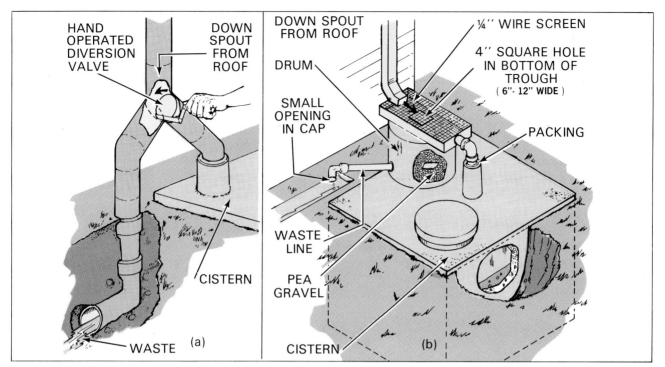

FIGURE 22. Methods of roof washing for cistern water. (a) Hand-operated diversion valve used to waste first rainfall. After roof is washed, the valve is changed so water will enter the cistern. (b) Automatic roofwash. The first rainfall flows into the drum. After the drum is filled, the remaining water flows into the cistern. During a period without rainfall, water dripping from the opening in the waste line empties the drum.

than the capacity of the filter, so much of it is wasted through the overflow pipe.

Cisterns can become **polluted from other sources;** for example, if the manhole cover does not fit well, if surface water can enter from the top, if a crack develops in the wall of the cistern and lets in ground water, or if tree roots force their way through mortar joints.

Recommendations for **cistern construction,** as given by the Environmental Protection Agency,[21] are as follows:

> Cisterns should be of watertight construction with smooth interior surfaces. Manhole or other covers should be tight to prevent the entrance of light, dust, surface water, insects, and animals.
>
> Manhole openings should have a watertight curb with edges projecting a minimum of 4 inches above the level of the surrounding surface. The edges of the manhole cover should overlap the curb and project downward a minimum of 2 inches. The covers should be provided with locks to minimize the danger of contamination and accidents.
>
> Provision can be made for diverting initial runoff from paved surfaces or roof tops before the runoff is allowed to enter the cistern.
>
> Inlet, outlet, and waste pipes should be effectively screened. Cistern drains and waste or sewer lines should not be connected.
>
> Underground cisterns can be built of brick or stone, although reinforced concrete is preferable. If used, brick or stone must be low in permeability and laid with full Portland cement mortar joints. Brick should be wet before laying. *High quality workmanship is required, and the use of unskilled labor for laying brick or stone is not advisable.* Two 1½-inch plaster coats of 1:3 Portland cement mortar on the

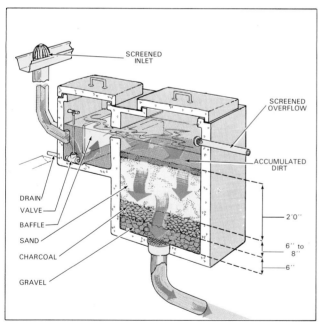

FIGURE 23. A sand filter of this type can be very effective in reducing pollution of cistern water if kept clean. If not cleaned regularly, it can be a source of pollution.

31

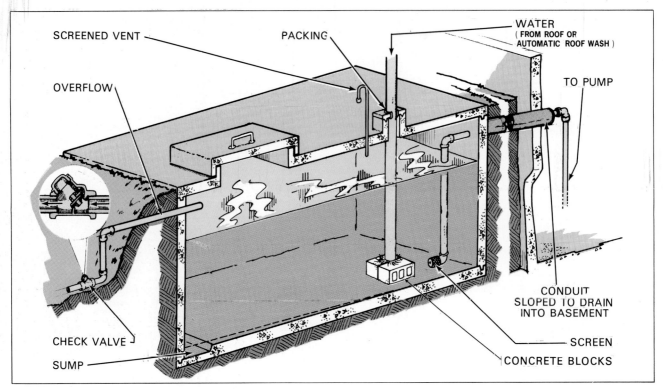

FIGURE 24. A cistern must be well constructed to keep surface water from entering around the points where piping enters the cistern, and where the manhole cover fits over the access hole. The check valve keeps rodents from entering the overflow pipe.

interior surface will aid in providing waterproofing. A hard impervious surface can be made by troweling the final coat before it is fully hardened.

Figure 24 shows how many of these precautions can be provided to protect your cistern from as much contamination as possible.

Another design is to build a primary and secondary basin. A cinder block or cement block partition divides the cistern into a primary and secondary basin. The rain water enters the primary basin and percolates through the partition. This "filtration" seems to work well.

In 1977, the National Interim Primary Drinking Water Regulations became effective.[38] You should become familiar with the provisions of these regulations.

REDUCING POLLUTION IN PONDS AND SMALL LAKES

If your water supply is to come from a **farm pond or lake,** it is impossible to avoid contamination. However, you can keep the contamination to a minimum by following these recommendations of the Environmental Protection Agency:[21]

Careful consideration of the location of the watershed and pond site reduces the possibility of chance contamination.

The watershed should:
1. Be clean, preferably grassed.
2. Be free from barns, septic tanks, privies, and soil-absorption fields.
3. Be effectively protected against erosion and drainage from livestock areas.
4. Be fenced to exclude livestock.

The pond should:
1. Be not less than 8 feet deep at deepest point.
2. Be designed to have the maximum possible water storage area over 3 feet in depth.
3. Be large enough to store at least one year's supply.
4. Be fenced to keep out livestock.
5. Be kept free of weeds, algae, and floating debris.

B. Determining if Your Water Supply is Safe

If your water source is a **well or spring,** and you have provided the pollution protection just recommended, your water supply is probably safe. But, for your own satisfaction and protection, it is well to have a sample checked occasionally. This is especially true if you notice any unexplained changes in taste, odor or appearance. Check with your state or local health department, or a local analytical laboratory for approved sizes and types of sample containers. Food and beverage jars and bottles cannot generally be cleaned adequately by the homeowner to prevent trace-level organic or metal contamination.

If you have an **open well,** or **one where surface water can enter directly,** there is no use having the water checked. It is almost certain to test unsafe. Follow the procedures outlined under "A Method of Protecting Water Sources from Pollution," then get a water test. Of particular significance is nitrate contamination from septic tanks and fertilizer application.

When your state or local health department tests a **water sample,** a check is made for certain kinds of bacteria called "coliform" bacteria. They are found in the intestinal tracts of warm-blooded animals. If that kind are present, it does not necessarily mean that your water sample has typhoid-fever germs or any other disease-producing bacteria. It is an indication that waste from humans or animals is entering your well or spring in some form. The waste may, from time to time, contain disease-producing organisms or parasites. You are taking a tremendous risk with your family if you use water polluted in this way without disinfecting it or finding the source of the pollution and correcting it.

Since **considerable care is required in collecting a water sample,** your state or local health office may wish to collect its own sample. If not, they will give you instructions for collecting a sample that will probably include the procedures shown in Figure 25.

If you have had a sample of well or spring **water checked and it was found to be unsafe,** check to make certain your well or spring is protected from *surface water* as outlined in the preceding discussion. If it is well protected, consider what underground sources around your area might be causing the pollution.

Since **underground contamination** often comes from such sources as cesspools, privies and leaky cisterns,

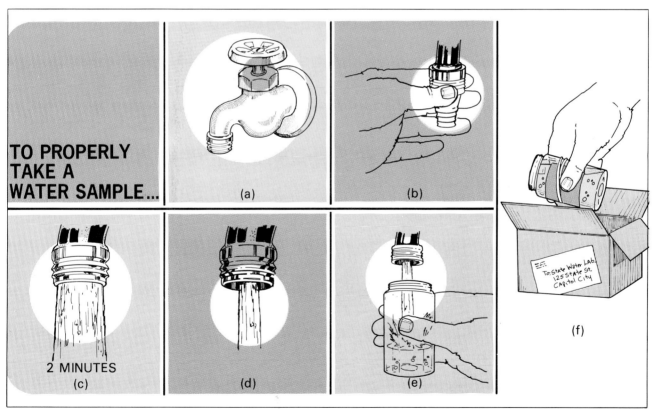

FIGURE 25. Collecting a water sample for testing involves several steps including the following: (a) select a coldwater faucet and check to make certain there is no leak around the packing gland, (b) remove any attachment, such as a hose connection, if one is present, (c) let the water run full flow for about 2 minutes, (d) reduce the water flow to about one third of full flow, (e) fill sterile bottle to within one-half inch of top and (f) mail or deliver the sample immediately to the laboratory.

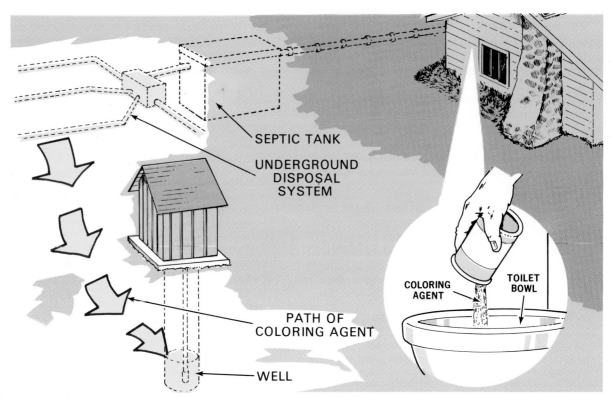

FIGURE 26. Use of coloring agent to check for underground contamination sources.

it is often possible to do something about it if you can find out for certain which one is causing the trouble. One method of checking is to empty *fluorescin* into the suspected source of contamination (Figure 26). If water from your well or spring becomes reddish or light green within a few hours, or even within a few days, you have found one source of contamination.

Fluorescin is strong enough as a coloring agent that it may be used whether the suspected contamination source is within a few hundred feet of your water supply or as far away as a mile. In strong solution it will be reddish in color. In weak solution, it is light green.

C. Methods Used for Disinfecting Water

In some areas there is no choice but to use water from polluted sources or from sources that may become polluted at times. As more people have had to face this condition, there has been increased interest in methods of disinfecting water to make certain it is safe for human use. Water for use by towns and cities has been disinfected for years, and the methods are well known. Disinfection methods for farm and home use are similar in many respects, but the equipment is usually more simple and less costly.

Most disinfecting units for small water supplies use **chlorine, heat,** or **ultra-violet light** for disinfecting water (Figure 27). All three are effective if the disinfecting units are properly cared for, if the water is clear, and if it contains little or no organic material. But many users are negligent, and water conditions vary widely. Because of this, opinions as to the most effective method of disinfection also vary widely—even among authorities. However, most public health authorities favor chlorine.

Another method for disinfecting water which is coming into prominence is the ozone treatment. Research is being done on this and other methods by contracts with USEPA. A unique method for producing small amounts of drinking water is by solar distillation.

In selecting any type of disinfecting unit, it is important that you seek the advice of your local health unit, or of the state health department, before making a purchase. Their experience with disinfecting units and with the water in your area, along with their knowledge of state and local regulations, can save you trouble and expense.

In order to understand their recommendations, it is important that you know something about how the various disinfecting units work. This information is presented under the following headings:
— Disinfecting by the chlorine method
— Disinfecting by the pasteurization method
— Disinfecting with the ultra-violet light method

DISINFECTING BY THE CHLORINE METHOD

The amount of chlorine needed to disinfect water is quite small—usually about 1 to 5 parts of chlorine to 1,000,000 parts of water. This is commonly referred to as **parts per million,** or "**ppm.**" If water contained no impurities, such as sulfur, iron, or organic particles from plants and animals that "use up" chlorine, the same amount could be used for treating all water supplies. But water is almost certain to contain one or more of these. For this reason, it is necessary to adjust the

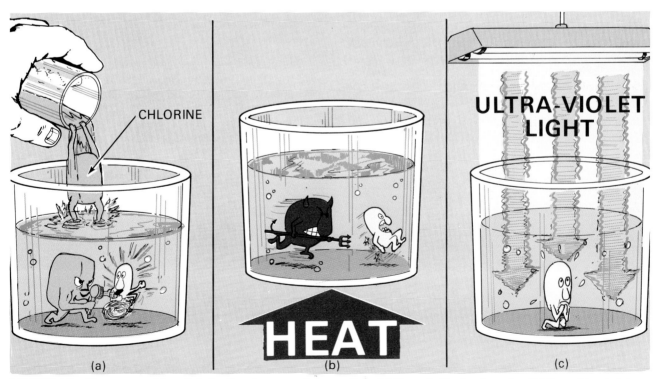

FIGURE 27. Means used for disinfecting home water supplies. (a) Use of chlorine in water. (b) Use of heat. (c) Use of ultra-violet light.

FIGURE 28. Chlorine for small-size chlorinators is available as: (a) laundry bleach (sodium hypochlorite), and also as (b) powder and (c) tablets (both calcium hypochlorite).

amount of chlorine to meet the needs of each water supply.

The **chlorine supply** can come from several sources. For home-size chlorination units the chlorine is available in the form of "hypochlorites"—diluted or low-grade forms of chlorine. There are two:

Sodium hypochlorite

Calcium hypochlorite

(Chlorine gas, the kind used to disinfect city water supplies, could be used, but it is too costly, too dangerous to handle, and requires too much attention for use with small water supplies.)

Sodium hypochlorite is a water-and-chlorine solution commonly used for laundry bleaches. It is available in two strengths—"domestic" or "commercial." *Domestic laundry bleach* is by far the most popular. It can be purchased from grocery stores under such trade names as Clorox, Purex, Hi-lex, Oxol, and No Worry. It contains only about *5.25 per cent of available chlorine*.

It is only the **available chlorine** portion that counts in figuring the parts per million. But even this small amount of chlorine treats a large quantity of water. For example, if you chlorinate at the rate of 1 ppm, one gallon of laundry bleach will treat about 50,000 gallons of water. If you treat at the rate of 5 ppm, one gallon of laundry bleach will treat about 10,000 gallons of water.

Commercial laundry bleach can be purchased from chemical supply houses and from some hardware stores in 5-gallon containers. It contains from *10 to 19 per cent available chlorine,* so smaller quantities of it can be used for treatment compared to domestic laundry bleach. For larger water systems, the commercial-strength bleach is usually more economical.

Some producers of hypochlorites are now including additional *cleaning agents* in their hypochlorite solutions. These are not considered a health problem, but they can cause the water to have a bad taste.

Calcium hypochlorite is available in powder and tablet form from local dairy-supply houses. It can be purchased under such trade names as B-K Powder, H.T.H., Perchloron, and Pittchlor (Figure 28b and c). It contains *30 to 75 per cent active chlorine by weight*. It is used to a very limited extent due to its tendency to form deposits that interfere with proper operation of the chlorinator unit. This is discussed further under "Care Required by Disinfecting Units," page 47.

There are a number of **chlorinator units** on the market, each of which is designed to meter small quantities of chlorine solution into a water supply as the water is being used. They are divided roughly into three types based on the means used to enter the chlorine solution into the water supply. They are as follows:

— Pump type
— Injector type
— Tablet type

The **pump type** is powered with an electric motor or water motor. These are often called "positive acting" or "positive displacement" types because they deliver a definite amount of chlorine solution with each stroke of the chlorinator pump. The pumping action is usually developed with a diaphragm as shown in Figure 29.

Here is *how it works*. As shown in the left inset, with the wobble plate in a retracted position, the diaphragm is drawn to the right by spring action. Suction is developed in the pump chamber. The intake valve opens, and chlorine solution enters the pump chamber.

In the right inset, the wobble plate has rotated to the extended position. The diaphragm is forced toward the pump chamber. This closes the intake valve and opens the outlet valve. The action causes a small amount of chlorine solution to be forced into the delivery line connecting the water pump to the pressure tank.

Provision for *adjusting the chlorine quantity* in pump-type units may be provided by: (1) varying the length of the pump stroke, (2) adjusting the pump speed, or (3) adjusting the amount of time the pump works, depending on the pump design.

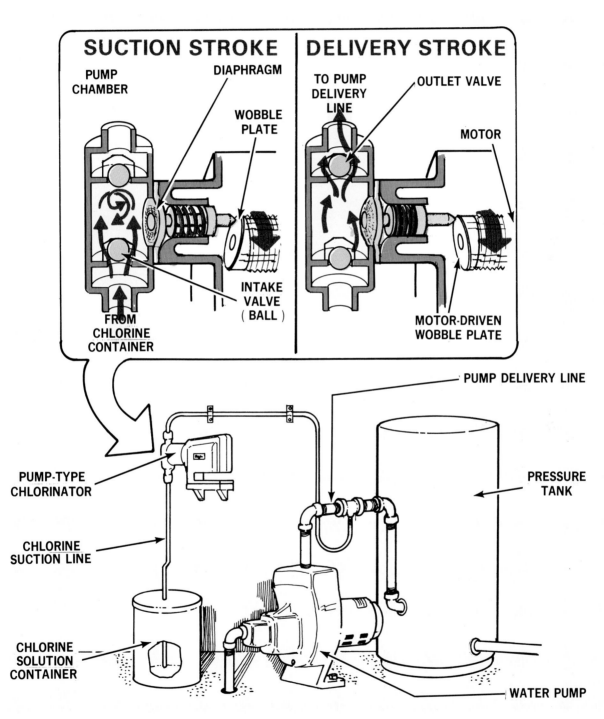

FIGURE 29. Pump-type diaphragm chlorinator driven by an electric motor. Chlorinator operates continuously while water pump is in operation. It is positive acting—delivering a fixed amount of chlorine solution with each discharge stroke.

The **injector-type chlorinator** is often called an "aspirator" or "jet." Figure 30 shows how it is connected to the water system.

Here is *how it works*. When the water pump starts, suction is developed in the water-pump suction line. At the same time, water from the pressure-tank supply line flows under pressure through the injector (inset) to the suction line. As water passes through the jet nozzle at high speed (velocity), suction develops in the chlorine-solution line. Chlorine is drawn into the jet stream and delivered to the water-pump suction line. The chlorine input is regulated by the feeding-adjusting screw.

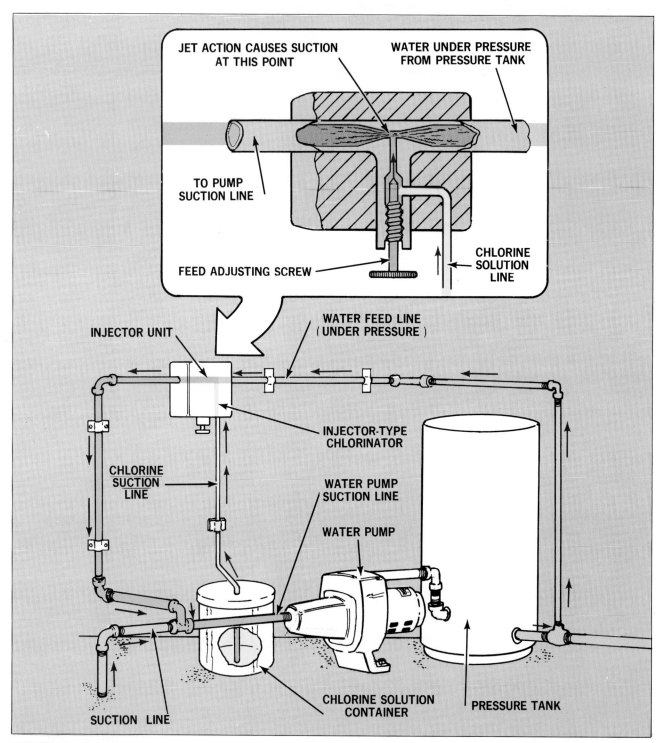

FIGURE 30. Injector-type chlorinator. Chlorine solution is drawn by jet (injector) action into the water-supply system.

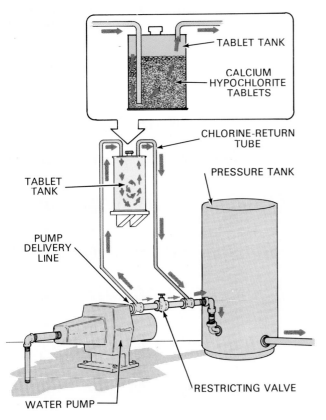

FIGURE 31. A tablet-type chlorinator bypasses water through a tank containing calcium hypochlorite tablets. The water dissolves a portion of the tablets and is returned to the pump delivery line ahead of the pressure tank.

With a **tablet-type chlorinator** (Figure 31), a small amount of water from the pump delivery line is circulated through a container of chlorine tablets and back into the delivery line. Water passing through the bed of tablets forms a chlorine solution.

The circulating action through the tablet tank is caused by the *restricting valve* in the pump delivery line. The restricting valve causes a slight increase in water pressure on the pump side of the restricting valve and a slight pressure decrease on the tank side. This is enough to cause some water to flow into the tablet tank and return through the chlorine-return tube to the pressure-tank side of the pump delivery line.

Commercial designs of the various types of chlorinators are shown in Figure 32.

DISINFECTING BY THE PASTEURIZATION METHOD

Water pasteurizers are a relatively new approach to disinfecting small water supplies. In general, the water pasteurizer works on the same principle, and at the same temperature, as is used for flash pasteurization of milk. The design was adapted to the disinfection of water by the Robert A. Taft Sanitary Engineering Center of the U. S. Public Health Service.

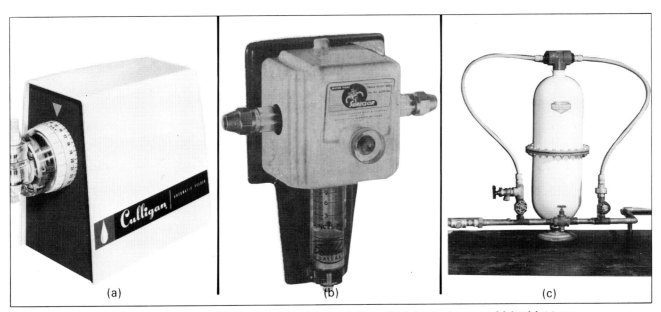

FIGURE 32. Commercial designs of chlorinator units: (a) pump type, (b) injector type, and (c) tablet type.

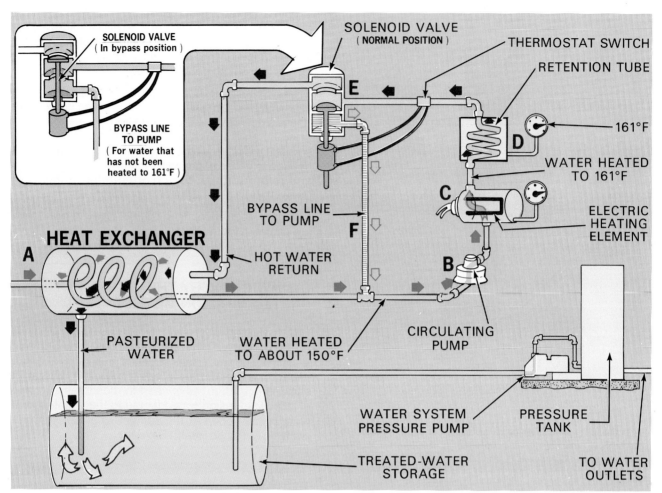

FIGURE 33. Pasteurizing type of disinfecting unit. Untreated water enters heat exchanger A where it is partially heated by water that has already been processed. Untreated water is moved through the circulating pump to the heating chamber C where the temperature is raised to 161° F. The temperature is maintained as the water passes through the retention tube D. The treated water then passes through the solenoid valve back to the heat exchanger where most of its heat is given up before it reaches the treated-water storage.

The **operating principle** is shown in Figure 33. Here is how it works. *Untreated water* enters the heat exchanger at A, where it is heated to about 150° F. by heat from the heated water that has already passed through the pasteurizing process. The preheated water is then drawn through the circulating pump B and forced through the heating chamber C. Here the temperature is raised to 161° F. and maintained at that temperature for at least 15 seconds as the water is forced on through the retention tube D.

If the *water is still holding at 161° F.* as it passes the thermostat, it moves through the solenoid valve E back to the heat exchanger, where it gives up heat to the cold incoming water. It is then discharged into the treated-water storage.

If the *water temperature is less than 161° F.* after passing through the retention tube, the thermostat switch activates the solenoid valve (inset) and the water is routed back through the by-pass line F to the circulating pump and returned to the heating chamber, where it is reheated until the proper temperature is reached.

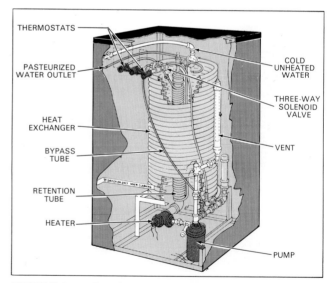

FIGURE 34. Sectional view of a commercially-built water-pasteurizing unit.

Since only a *small amount of water can be treated* at one time, the treated water is stored in a tank and pumped by a second pump to the various water-use outlets.

For the unit to *operate economically,* it is necessary that it operate at least half time. For that reason, the *capacity on present units is limited to approximately 0.35 to 0.4 gallons per minute (20 to 24 gallons per hour).* If it were to operate for less time, there would be more heating-up and cooling-off periods. This greatly reduces its efficiency in the use of heat.

DISINFECTING WITH THE ULTRA-VIOLET LIGHT METHOD

Disinfection of water by means of **ultra-violet light** has been in use for a number of years. It consists of one or more ultra-violet lamps each encased in a quartz sleeve. A thin layer of water is passed around each sleeve to expose it to ultra-violet light. The killing action is the same as that provided by sunlight in killing bacteria in open streams. The ultra-violet lamps are designed to provide a maximum amount of light at the proper wave length for the greatest bacteria-killing action—2,537 Angstrom units. (The Angstrom is a measure of light wave lengths. One centimeter—about .4 inch—is equal to 100,000,000 Angstroms.)

Figure 35 shows the operating principle and the control units for two different types,

For an ultra-violet disinfecting unit to be effective, the **water must be circulated** in such a manner as to expose each water droplet to as much light as possible. The killing action is fast if the light can reach the bacteria.

If your water supply contains small dirt particles, you will need to install a filter to remove them if an ultra-violet unit is to do an effective job. Otherwise, some bacteria or virus may pass through the unit in the shadow of dirt particles without being reached by the killing rays of the ultra-violet light.

The purpose of a *quartz sleeve* around each lamp is to protect the lamp from the cooling action of the water. The lamps must maintain a certain level of heat to produce the necessary killing effect.

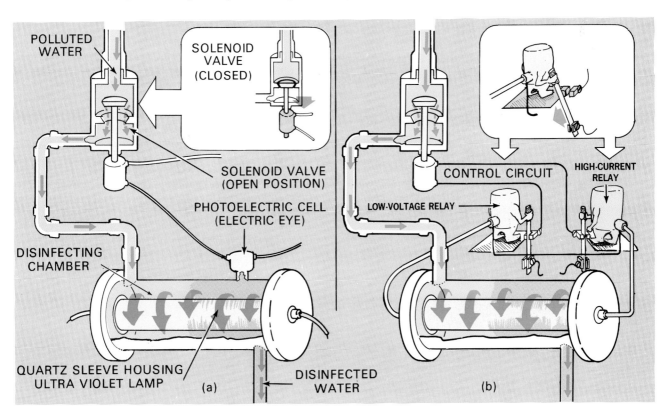

FIGURE 35. Types of automatic safety controls used for ultra-violet disinfection units. (a) Photo-electric cell holds the solenoid water valve open as long as the water passing through the disinfecting chamber is exposed to sufficient ultra-violet light. If dirt collects on the tube, or the tube breaks, or age limits the amount of light output, or there is a power outage, the photo-electric cell closes the solenoid valve (inset). (b) Another type of ultra-violet control system provides protection with low-current and high-current relays. In case of no current or low-voltage, the low-voltage relay opens the control circuit (inset), which causes the solenoid valve to close and shut off the water supply. As the lamp grows older and decreases in light intensity, the current (amperes) increases and reaches a point where the high-current relay opens the control circuit and shuts off the water supply by means of the solenoid valve.

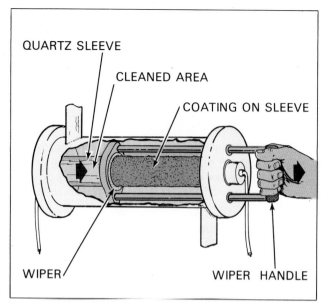

FIGURE 36. Hand-operated wiper for removing coating that forms on the quartz sleeve.

Each unit is *designed for a certain maximum flow* of water. Some are equipped with a flow-regulator valve so the amount of water can be adjusted at the unit to the capacity of the unit. There is a wide selection of capacities available to fit different home and farm needs. The sizes vary in capacity from about 22 gallons per hour to 20,000 gallons per hour.

The *intensity of the ultra-violet light given off by the lamp* gradually decreases as the lamp is used. It will finally reach a stage where it is relatively ineffective for killing bacteria. To avoid this situation, some units are equipped with an "electric eye" (photo-electric cell) which is sensitive to the amount of ultra-violet light given off by the lamp(s) (Figure 35a). This device measures the amount of ultra-violet light that has passed through the water. When the intensity gets too low, the electric eye shuts off the water supply automatically (Figure 35a, inset) until a new lamp is installed. Cloudiness of the water, a film or coating on the quartz tube around the lamp, or a power failure will also cause the solenoid valve to close.

Another safety device provides a *time delay* after the lamp is turned on. This allows the lamp to reach its greatest output of ultra-violet light before the solenoid valve can open.

Still another type of unit uses two electrical relays to provide safety against improper operation. They are sensitive to the amount of current being supplied to the lamp(s). If the *voltage is too low* to supply enough current to the lamp for proper ultra-violet intensity, the low-voltage relay acts to shut off the water supply (Figure 35b).

As the *lamp gets older* and the ultra-violet light intensity drops off, the current supply to the lamp gradually increases. When it reaches a certain maximum, the high-current relay acts to turn off the water supply.

Some ultra-violet units have *no automatic protection*. With this type, the manufacturer usually recommends changing lamps at the end of six months or some other fixed period of time.

The *quartz sleeves that protect the ultra-violet tubes tend to become coated* with small particles that lower the ultra-violet light intensity. Consequently, manufacturers provide wipers for cleaning the sleeves without removing them. Some are operated by hand (Figure 36). Others provide automatic wiping of the sleeves to keep them free of deposits. A small water turbine operates the wipers whenever water flows through the unit.

Ultra-violet units are *designed for continuous operation*. They are usually installed on the outlet side of the pressure tank or close to the faucet that supplies drinking water.

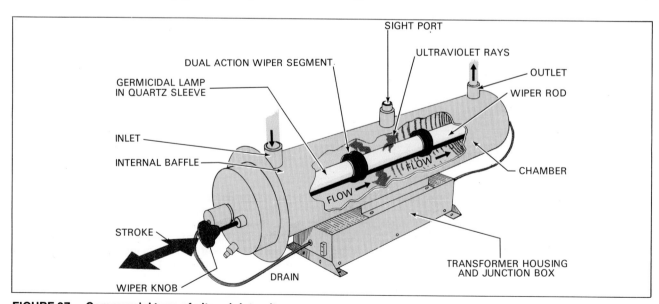

FIGURE 37. Commercial type of ultra-violet unit.

D. What Disinfecting Method to Use

Table V lists most of the factors you need to consider in selecting a disinfecting unit. It also compares the effectiveness of each when used to disinfect small water supplies—at least as nearly as effectiveness can be determined at this time. As mentioned earlier, authorities do not agree on the effectiveness of each method.

In comparing effectiveness of the three methods of disinfecting it is assumed that whatever disinfecting unit is used will be **properly maintained and checked regularly** to make certain it is working satisfactorily. If this is not done, you may not get the disinfecting action you wish. It is around this situation that considerable controversy centers as to the merits of each method of disinfection.

In order to understand Table V, you will need to know the meaning of such terms as "dosage," "demand," "residual," and "superchlorination," and something about the different ways chlorine is applied to water. There is also another factor involving the care required. This information is included under the following headings:

— Effectiveness of chlorine
— Care required by disinfecting units

EFFECTIVENESS OF CHLORINE

It takes a very small quantity of chlorine **to kill bacteria.** Somewhat more is needed **to kill viruses** quickly but, whatever quantity is needed, it is still so small that it is measured in "parts per million" (ppm). As explained earlier, one ppm is the same as adding one gallon of available chlorine to one million gallons of water, or one pound of available chlorine to one million pounds of water. (You may also see the expression "milligrams per liter"—mg/l. For water, parts per million and milligrams per liter are approximately equal.)

The total amount of chlorine added to water—in parts per million—is called *"dosage."* As quickly as a chlorine solution is added to most water supplies, a portion is "used up." It reacts with particles of organic matter, if there are any in the water, and with certain minerals, slime, or chemicals that may be present. Chlorine "used up" in this manner is called *"chlorine demand."* If only enough chlorine is added to meet this need, there would be little or no disinfecting action.

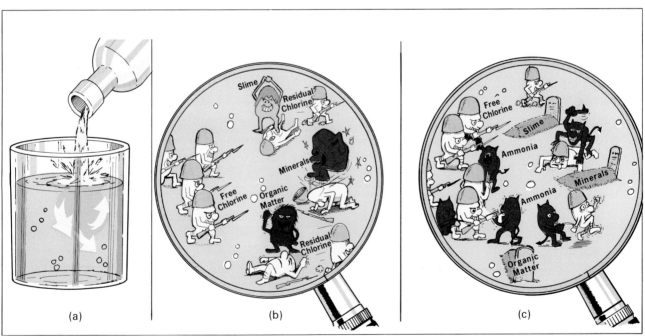

FIGURE 38. (a) Dosage is the amount of chlorine added to water. (b) Organic matter and minerals will "use up" some of the chlorine causing it to lose its killing action. Any chlorine that is left over is called "residual chlorine." Residual chlorine may be "free," as shown or (c) if ammonia is present (as in most pond water), some chlorine will combine with it. Combined chlorine is much slower in its killing action than free chlorine because of this handicap.

TABLE V. COMPARISON OF WATER DISINFECTION METHODS

Effectiveness Under Normal Conditions:	Disinfection Methods			
	SIMPLE CHLORINATION	SUPERCHLORINATION-DECHLORINATION	ELECTRIC PASTEURIZATION	ULTRA-VIOLET RADIATION
—FOR KILLING BACTERIA	Effective	Effective	Effective	Effective for clear water
—FOR KILLING VIRUSES	Effective with some viruses. Complete effectiveness has not been determined.	Effective with many viruses. Complete effectiveness has not been determined.	Effective with many viruses. Complete effectiveness has not been determined.	Effective with many viruses. Complete effectiveness has not been determined.
SPEED OF KILL (Bacteria)	Requires at least 20 minutes contact time with minimum chlorine residual of 0.2 to 0.5 ppm.	For killing virus—30 minute contact time with minimum chlorine residual of at least 0.5 mg/L at pH below 8.0 is required.	15 seconds	Fast acting at proper light intensity level in water that is free of suspended particles and ultra-violet absorbing matter in solution.
Other Factors To Consider:				
EFFECT OF MINERALS IN WATER	Some chlorine is "used up" if iron or sulfur is present in water. If mineral content varies from time to time, dosage will need to be adjusted with simple chlorination to maintain proper chlorine residual. With super-chlorination, dosage is not readily affected. In waters of low total solids content, the chlorination chemical may greatly affect the acidity or basicity of the water (i.e. inadequate buffer capacity), and may increase ability of the water to dissolve pipe and solder. This is much more of a problem when acidic chlorine solutions are used.		Heating may cause mineral deposits to form thus slowing heat movement.	Minerals gradually coat lamp sleeve surfaces and reduce efficiency.
EFFECT OF HIGH WATER ALKALINITY	Purifying action slowed.	Purifying action slowed.	Not affected	May tend to coat lamp sleeve(s).
EFFECT OF SUSPENDED PARTICLES IN WATER (Such as contained in pond water)	Water should be filtered to remove particles. Otherwise, it is difficult to maintain proper amount of chlorine residual, and the particles may protect some bacteria from the killing action of the chlorine.		Slows heat movement. May foul heat exchanger.	Greatly reduces effectiveness. Particles may protect some bacteria from killing action of light. Water must be effectively filtered first to remove all particles.
EFFECT OF INCOMING WATER TEMPERATURE	Increase in temperature speeds disinfecting action. Lower temperatures slow action.		Purification is not affected.	Most efficient at 100° F. water temperature. Less effective as temperature lowers.
RESIDUAL EFFECT (Ability after treatment to keep water disinfected)	With chlorine residual of 0.2 ppm or more, protection continues for several hours after treatment.	3 ppm or more residual provides excellent protection for many hours after treatment.	No protection after leaving pasteurizer.	No protection after leaving ultra-violet unit.
EFFECT ON WATER TASTE	May have some chlorine taste but is still palatable.	Not palatable for some humans until dechlorinated. Activated carbon filter is used at kitchen faucet to remove all chlorine taste for drinking-water purposes. Water is palatable to livestock and poultry.	Taste not affected	Taste not affected
PROTECTIVE MEANS USED TO ASSURE PROPER OPERATION	Color check with test kit enables user to determine amount of residual chlorine present. Water should be checked weekly.	Odor of chlorine is noticeable in superchlorinated water before dechlorination.	Solenoid (electric) valve shuts off water supply when heating element burns out. It also returns water to heater if inadequately heated.	Equipped with solenoid (electric) valve to shut off water supply when lamp(s) dims or burns out, or electric service is interrupted.
CAPACITY	Available for any capacity water system	Available for any capacity water system	20 gal. per hour (Size presently available)	Various size units available for any capacity water system.
ADVANTAGES IN ADDITION TO DISINFECTION	Can be used to remove iron, sulfur, and certain tastes and odors. Kills iron and sulfur bacteria.		None	None

If the dosage is more than enough to take care of the chlorine demand, any that is left over is called *"chlorine residual."* It is this portion that provides the disinfecting action. However, if *ammonia* is present in the water, as is usually the case with pond or surface water, a portion of the chlorine residual will combine with the ammonia to form what is called *"combined chlorine residual."* That which has not combined is called *"free chlorine residual."* Over the same period of time, the free chlorine residual is up to 25 times more effective in destroying bacteria than the combined chlorine residual. Figure 38 shows an animated concept of what happens to chlorine in water.

Two reasons why **health authorities, in general, favor chlorine disinfection** over other disinfection methods discussed here are: (a) the chlorine residual lasts for a long period of time after leaving the disinfecting unit thus *providing continuing protection,* and (b) you can measure the amount of chlorine residual with a test kit so at any time you can *determine how much protection* is being provided. (Use of the test kit is discussed later.)

As a result of these varying water conditions, two different techniques for continuous chlorination of small water supplies are used. They are known as: (1) simple chlorination and (2) superchlorination.

SIMPLE CHLORINATION—Simple chlorination follows closely the standards used by towns and cities for disinfecting their water supplies. Enough chlorine is added to the water to meet the chlorine demand (normally, about 0.5 to 1 ppm), plus enough additional to supply about **0.2 to 0.5 parts per million residual** when checked after 30 minutes of contact time.[21] Most states have their own regulations on minimum residual and minimum contact time, which may differ slightly from this. This is enough chlorine and enough contact time to kill bacteria.

Simple chlorination may not be enough to kill certain of the **viruses.** They are much more resistant than bacteria. Chlorine as a disinfectant **increases in effectiveness:** (1) as the chlorine residual is increased, and (2) as contact time is increased.

Chlorine **becomes less effective:** (1) as water temperature lowers, and (2) as it increases in alkalinity. As you probably recall from your science courses, the degree of acidity or basicity is expressed in "pH." Water with a pH of 7 is neutral—neither alkaline nor acid. The smaller the pH number, below 7, the more acid the water. The higher the number, above 7, the greater the alkalinity of the water. If water has a pH of 5 to 6, a chlorine residual of 0.2 ppm may be satisfactory. If the pH is 7.5 to 8, a residual of 0.8 ppm may be needed.

The **residual** is determined with home-size chlorinators by means of a test kit which is supplied with the chlorinator unit, or it can be purchased separately. The test is made after the *chlorine has had at least 20 minutes of contact time.* This is discussed more completely under the next heading.

With this amount of residual chlorine you may detect a slight *chlorine taste,* but the water is not objectionable for drinking purposes.

The problem of how to provide at least **30 minutes of contact (retention) time** before the water is used is a difficult one with small water systems. At one time it was thought that the water-system pressure (storage) tank would help provide retention time. But tests[22,23] show that water "short circuits" or channels through a pressure tank so it is not a dependable method of providing even a portion of the contact time. A large size pressure tank is not the answer either.

One method of providing over **30 minutes of retention time** is to use a *gravity tank* (Figure 84a), or *reservoir* (Figure 84b), that will hold enough water to provide ample contact time. If you are disinfecting water from a pond, you are almost certain to have this type of reservoir already installed. It is the one that accumulates water that is filtered from the pond. The reservoir is large enough to provide ample water to meet the normal needs of the home and farm. Usually it provides enough capacity for ample chlorine contact time.

Another method is to **feed the chlorine solution into your well** at a point near the pump intake (Figure 39a). When the pump starts, the chlorinator starts automatically. Chlorine solution enters the water in the well and is drawn into the pump suction along with the well water. The end of the tube must be close to the end of the pump intake for this arrangement to be effective. The flow of chlorine solution stops when the pump stops. This arrangement helps provide more contact time.

A popular method is to use an automatic chlorine pellet dispenser (Figure 39b). It feeds the pellets directly into the well casing at a predetermined rate, depending on your water use. The amount recommended is 0.4 to 0.8 ppm for human consumption and up to 3 ppm for livestock.

SUPERCHLORINATION—Superchlorination is a technique used to **shorten the contact time** needed for disinfection. This is important if you have no reservoir or other means for providing the 30-minute contact time needed for simple chlorination.

With superchlorination you substantially increase the amount of chlorine added to the water. *This greatly reduces the time needed to kill bacteria.*

Authorities are not in complete agreement on recommendations as to how much chlorine should be added for superchlorination, or how much contact time to allow to kill bacteria. The U.S. Environmental Protection Agency (EPA)[21] recommends a chlorine residual of not less than **3 ppm and a contact period of 5 minutes.** One study indicates that by raising the chlorine residual to 5 ppm, less than 10 seconds are required for effective bacterial kill even under unfavorable conditions.[24]

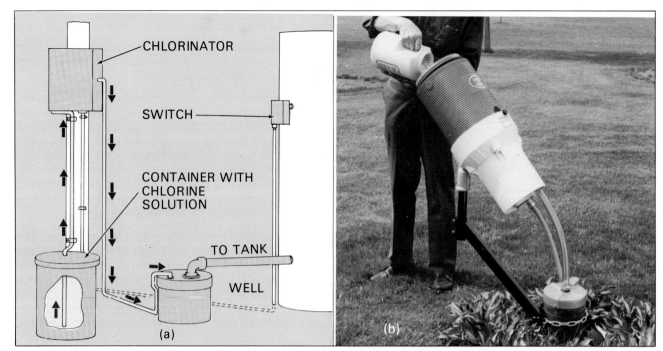

FIGURE 39. (a) One method of increasing chlorine contact time is to feed the chlorine solution into the well near the pump intake. (b) A chlorine pellet dispenser may be installed that can be adjusted to feed the correct amount of chlorine as the water is used.

Iowa studies[23] show that if you increase the amount of chlorine until you have a free **chlorine residual of 5 to 6 ppm** and a contact time of **7 minutes,** you can **kill both bacteria and viruses** under the worst conditions that you are likely to encounter.

If you decide to use superchlorination, and do not have a storage tank that will **provide the contact time,** you may have enough *supply-pipe length* between your water pump and your first water outlet to provide the contact time. Table VI shows the pipeline lengths and pipe sizes needed to provide the contact time at various pump capacities.

For adequate mixing of the chlorine and water, some turbulence is required for mixing as the water flows through the pipe. This is taken into account in the table.

If there is only a *short distance* between your water pump and your nearest water outlet, you can coil the extra length of plastic or copper pipe that is needed into a compact roll near the pressure tank (Figure 40a).

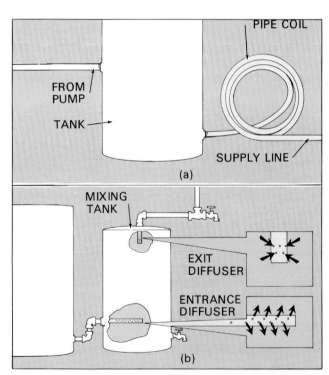

FIGURE 40. Ways of providing enough contact time for chlorine to do an effective disinfecting job. (a) Addition of a pipe coil in the supply line from the pump. (b) Providing a special mixing tank in the supply line designed to prevent channeling of water.

TABLE VI. MINIMUM LENGTH AND MAXIMUM SIZE OF PIPE NEEDED TO PROVIDE 7 MINUTES CONTACT TIME AT FULL PUMP CAPACITY (approximately)

Pump Capacity	Pipe Sizes (inches)					
	1	1¼	1½	2	2½	3
(gal. per hr.)	Pipe-line Lengths (ft.)					
250	935	605	425	240	150	110
300	1130	730	510	285	185	115
350		855	600	332	215	135
400		980	685	380	245	150
450		1085	760	420	270	170
500			845	480	305	185
550			935	520	335	215
600			1020	565	356	225

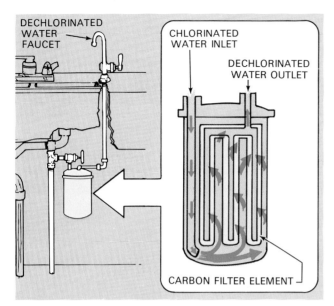

FIGURE 41. Precoat carbon filter. This type is frequently used to remove chlorine from drinking water that has been superchlorinated. The filter element can be removed and replaced with a new one when needed.

The USDA has developed a **mixing tank** which will provide the necessary contact time (Figure 40b). It is designed so the water does not channel, as it does in the standard pressure storage tank. By providing one that is 10 to 15 times larger than the pumping rate of your pump, in gallons per minute, it is adequate to provide ample contact time.

If you are interested in a tank of this type, the plans for constructing it are available from your state Cooperative Extension Service. Ask for plan number 6155, Water Conditioning Tank, USDA.

If you find the **chlorine taste** in superchlorinated water is too strong for drinking purposes, the chlorine can be removed by running the drinking water through activated carbon. Figure 41 shows a small precoat carbon filter that can be installed in the water line near the outlet where drinking water is supplied. It will remove all of the chlorine from the water.

Livestock and poultry, after a brief period, adjust to superchlorinated water and even prefer it. At the same time, it is very effective for controlling water-borne diseases.

If you are interested in using superchlorination, get recommendations from your county sanitarian.

CARE REQUIRED BY DISINFECTING UNITS

Table VII lists the more important care and maintenance needs for the different types of disinfecting units. However, some additional explanation is needed for the chlorine units. This is included in the following discussion.

Each manufacturer supplies instructions as to how the **chlorine solution** should be mixed.

If you are using the **simple-chlorination method,** the instructions will probably indicate that you should dilute the laundry bleach with several parts water to one part bleach. If you are using calcium hypochlorite, which contains a much higher precentage of chlorine, much more dilution will be required.

If you are using the **superchlorination method** of disinfecting water, the instructions may indicate that the chlorine bleach solution be used without dilution.

Most manufacturers recommend the use of laundry bleach diluted with *soft water.* Sometimes distilled or demineralized water is used. This is important. If you dilute with hard water, hardness deposits are almost certain to form on the parts of the equipment that are in contact with the solution. As the deposits build up around the valve parts and small openings, they gradually interfere with operation until the unit may cease to function. It is then necessary that you completely disassemble the unit and clean out the deposits.

If *hard water* is all you have available, you can add a softening agent to the water in the solution container.

Chlorine solutions will gradually *lose their strength.* Consequently, it is generally recommended that new solutions be mixed every month or two. The solution will hold its strength longer if it is kept in a cool place and out of direct sunlight.

Chlorine is **highly corrosive** and will react with metal containers. Consequently, the solution container should be of glass, plastic or earthenware.

If you have difficulty with **sludge and undissolved materials** collecting in the bottom of the solution container, you can secure a foot-valve assembly (Figure 42), which protects the foot valve from sediment that collects on the bottom of the solution container.

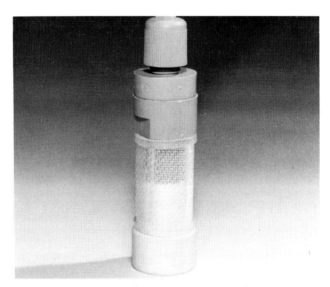

FIGURE 42. A chlorinator foot-valve assembly designed to protect the foot-valve from sediment and sludge that collect on the bottom of the chlorine-solution container.

TABLE VII. CARE AND MAINTENANCE REQUIRED FOR DISINFECTION UNITS

Service and Inspection Jobs	Methods of Disinfection			
	SIMPLE CHLORINATION	**SUPERCHLORINATION-DECHLORINATION**	**ELECTRIC PASTEURIZATION**	**ULTRA-VIOLET**
MIXING OF CHEMICAL SOLUTION	Laundry bleach is diluted with soft water. Proportions vary, 1 part laundry bleach diluted with 4 to 10 parts water. Mix every 2 months in cool weather, monthly in hot weather. Amount of chlorine needed—approx. 1 gal. of laundry bleach (5.25% solution) will treat 50,000 gal. of water at rate of 1 ppm.	Can use laundry bleach (sodium hypochlorite) without dilution. Calcium hypochlorite requires dilution. Amount of chlorine needed—one gallon of laundry bleach treats about 10,000 gallons of water at rate of 5 ppm.	No chemicals used	No chemicals used
MECHANICAL OR ELECTRICAL SERVICING	**Footvalve** on chlorinator may stick from accumulation of deposits if hard water or calcium hypochlorite is used. **Valve sticking** may develop on **pump-type units** for same reason. **Orifice clogs** on jet-type units. (The orifice is the small opening in the jet which develops pumping action.)		Heating element(s) may need to be changed about every 3 to 5 years. (Field experience is too limited to know whether other troubles may be encountered.)	Units with automatic safety controls require lamp tube replacement about every 1 to 3 years. If operated without automatic safety controls, change lamps about every 6 months. Tubes may break in plumbing systems that develop water hammer.
RECOMMENDED FREQUENCY OF INSPECTION	Weekly	Weekly	Weekly	Weekly
RECOMMENDED FREQUENCY OF CLEANING	Twice yearly to check and remove any accumulated deposits (and clean strainers on some units).		No definite period; however, accumulating tank may have to be cleaned or disinfected with chlorine occasionally. May have to flush deposits from pasteurizer yearly with Calgon or similar cleaner.	Tubes should be removed and cleaned about every 4 to 6 months. Filter, used to remove small particles, should be serviced monthly.

If you are likely to forget and let your **supply of chlorine solution run low,** you can get a device that will sound an alarm when this happens. It can be connected so it will shut off your water pump and sound a bell. Before the pump will start, you will need to refill the chlorine-solution container.

It is important that you **check the chlorine residual regularly.** Some authorities recommend a daily check, others a weekly check, particularly if the amount of contamination in your water supply varies considerably. If it is fairly constant, monthly checks may be adequate.

The **chlorine-residual test** may be made with Palin DPD. When added to water containing chlorine, it causes a greenish-yellow color to develop. The intensity of the color is an indication of the amount of residual chlorine present. You can check the color that develops against a standard color chart which gives you an approximation of the amount of residual chlorine present.

To **make the test,** a measured sample of water is placed in a glass tube. A certain amount of orthotolidine is added in accordance with the instructions that accompany the test kit.

The yellow color that develops in the first 15 seconds is an indication of the **free available chlorine**. The **combined available chlorine** reacts more slowly and requires about five minutes at 70° F. for full color development. As pointed out earlier, it is the free available chlorine that is by far the most active disinfectant. So, in comparing the color at the end of the 15-second period, it is important that the color be intense enough to show the presence of ample free chlorine.

The **accuracy of this test** is dependent on how accurately the water is measured, the freshness of the orthotolidine—it should not be kept more than 6 months—and the ability of the person to distinguish among the shades of yellow.

Mechanical servicing consists mostly of freeing stuck valves and unclogging passages. If the chlorine solution you are feeding through your chlorinator is free of deposits, you will get away from most of the difficulty. If the chlorinator becomes clogged and must be taken apart, this can lead to another difficulty. The parts of the pump that are in contact with chlorine are commonly made of plastic. When they are assembled and tightened with a wrench, the parts may be ruined. Most manuals recommend that the plastic units be tightened by hand.

FIGURE 43. Orthotolidine Test kit used to determine the amount of residual chlorine in water.

E. Providing Protection Against Radioactive Fallout

In case of a nuclear attack, safe water—that protected from radioactive fallout—becomes important. At the time of an atomic burst, there is a large amount of radioactive dust in the mushroom cloud that develops. The dust scatters over thousands of miles and gradually falls back to earth (Figure 44).

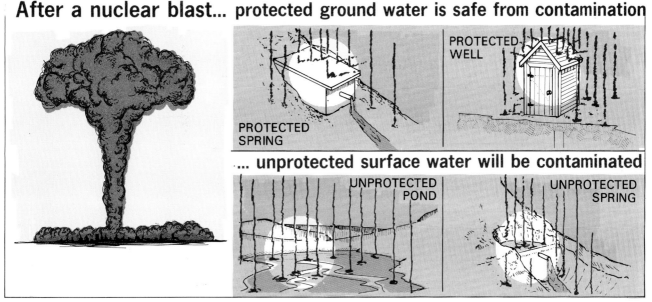

FIGURE 44. In a nuclear explosion, tremendous quantities of dust are taken up into the mushroom-shaped cloud. This becomes radioactive and gradually falls back to earth. Ground water supplies, if properly protected against general contamination, are protected from radioactive contamination. Open dug wells, ponds or unprotected springs will become contaminated with radioactive dust.

Open water supplies, such as ponds, lakes or streams, are readily contaminated.

Ground water exposed to the atmosphere, such as in an open dug well or spring, soon becomes contaminated unless protected by a cover. Any type of cover that would normally keep out dust is considered satisfactory. If a spring or well is not already covered, you can use a tarpaulin to catch the radioactive dust and then remove the tarpaulin after the danger period has passed.

Underground water supplies that are covered and well protected from surface water are *protected from radioactive fallout*. As water percolates through the soil, radioactive dust is removed by the first few inches of top soil. Although sand is not as effective as clay, it is still very effective—the water must percolate a few inches further before the radioactive dust is removed.[25]

Other sources of radioactivity such as pollutants and natural minerals are possible, also.

To some extent, a household **water softener** will remove radioactive material, but its capacity is so limited it soon becomes overloaded.

III. Determining the Need for Water Conditioning

Now that you have made provisions for a water supply that is safe—free from disease-causing bacteria, viruses, and parasites—you may still have one or more checks to make. Your water supply may contain **minerals, acids, or foreign matter,** which keep it from being fully satisfactory for your needs. In this discussion, correcting water with these quality problems is called "water conditioning."

Perhaps you have wondered why some water requires so much soap to make it suds. You may have seen water where **red particles** form in a freshly-drawn glass of water and settle to the bottom. Or you may have noticed **black particles** develop. You may have wondered about the **peculiar odor or taste** of some water. Other water-quality problems may show up in the form of **corrosion**—such as pitted pipes or a buildup of **scale** inside of your piping system. These are water-conditioning problems.

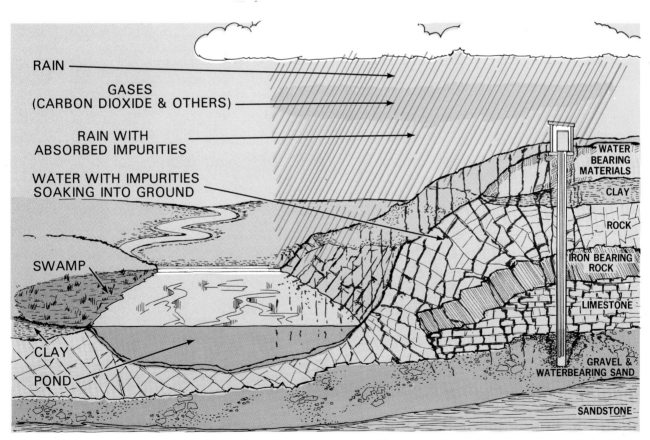

FIGURE 45. Rain water is usually considered fairly pure, but as the droplets fall through the air, they absorb some impurities. Gases, such as carbon dioxide, join with water to form weak acids. As the rain water filters through the earth's surface, the acids react with the minerals to change the character of the water, thus changing its quality.

51

The reason water develops these qualities is because it combines readily with some chemicals to make it **chemically active.** For example, rain falling through the air combines with carbon dioxide in the air to form a very weak acid called "carbonic acid." As the carbonic-acid water filters through sand, soil, and rocks, it dissolves small amounts of calcium, magnesium, iron, and other minerals and holds them in solution (Figure 45). When this happens, the quality of the water changes.

As a result of many possibilities for the pollution of water, the U.S. Government has authorized the EPA to regulate certain water facilities. This was authorized in 1977 through the National Interim Primary and Secondary Drinking Water Regulations.

As water flows through swamps or collects in ponds and lakes, it takes up **flavors and odors** from plants and decaying vegetable matter.

Most of these water problems can be overcome. Your job is to find out the cause and extent of your water-conditioning problem(s) and determine what means to use in correcting them. Use Table VIII for this purpose.

There are some explanations in the table that need to be expanded further for clarity. They are included under the following headings:
— Determining the probable cause of poor water quality
— Measuring the extent of the water-quality problem
— Methods of correcting the water-quality problem

DETERMINING THE PROBABLE CAUSE OF POOR WATER QUALITY

Table VIII, column 1, lists **symptoms** that describe various conditions of water supplies.

Column 2 indicates the probable **cause of a symptom.** You may need more information than is contained in the table to better understand the cause of the condition. This discussion is to provide the additional information.

HARDNESS—If you found that the symptoms indicate water hardness, as indicated in column 1, it is probably due to the presence of **calcium and magnesium** in the water. Aluminum and iron will also contribute to hardness, but aluminum is usually present in such small quantities that it has little effect on hardness. Control of iron is discussed next. All of these minerals are dissolved by the water as it passes through soil and over rock formations. For this reason, surface water is usually softer than water from wells. It has had less opportunity to contact soil particles and rock formations containing minerals.

RED WATER—If you found symptoms of "red water" (iron), it could have developed in two different ways:
— By the **dissolving action** of water as it percolates through *underground iron deposits* (Figure 46a). Or, by dissolving action as it remains in *contact*

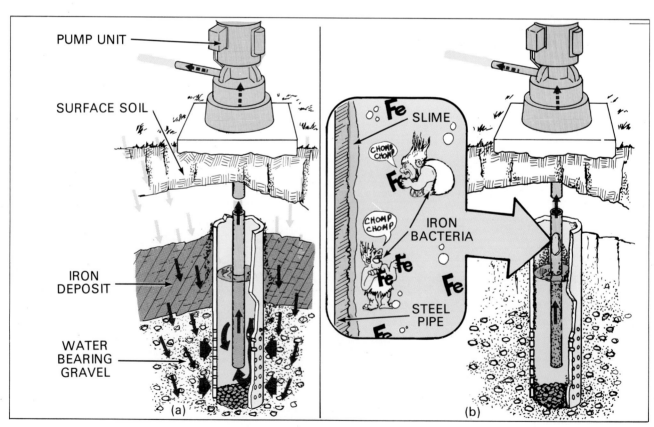

FIGURE 46. (a) Most of the iron present in water has been picked up as the water passes through underground iron deposits. (b) Iron bacteria work on the iron in the water to form a red slime.

TABLE VIII. ANALYZING AND MEASURING WATER-CONDITIONING PROBLEMS

Symptoms (Col. 1)	Probable Cause (Col. 2)	Measurement (Col. 3)	Method(s) of Correction (Col. 4)
HARDNESS —Sticky curd forms when soap is added to water. Causes well-recognized ring in bathtub. —The harder the water, the more soap required to form suds. —Glassware appears streaked and murky. —Hard, scaly deposits form inside of metal pipes. —Your skin roughens from washing.	Calcium and magnesium in the water (may be in the form of bicarbonates, sulfates, or chlorides). Iron also contributes to hardness. (See next group of symptoms—Red Water.)	Test kit: Standard soap solution is added to two oz. of water one drop at a time. Solution is shaken until lasting suds form. Number of drops are approximately equal to grains per gallon of hardness. (One grain is equal to 17.1 ppm.) 0- 3 drops (approx. 50 ppm) Soft water 4- 6 drops (50-100 ppm) Moderately hard 7-12 drops (100-200 ppm) Hard 13-18 drops (approx. 200-300 ppm) Very hard 19 or more drops (over 300 ppm) Extremely hard	—**3 grains** (approx. 50 ppm) hardness, or less, no softening needed. —**More than 3 grains** hardness, use a zeolite (ion-exchange) softener or reverse-osmosis unit.
RED WATER **Dissolved Iron** —Red stains appear on clothes and porcelain plumbing fixtures, even if as little as 0.3 ppm is present. —Causes corrosion of steel pipes. —Water has metallic taste. —Freshly-drawn water sometimes appears clear at first. After exposure to air, rust particles form and settle to bottom of container.	Iron (sometimes including manganese) Caused by dissolving action of water as it passes through underground iron deposits, or contacts iron and steel surfaces.	Test kit: —Standard acid solution is added to water sample to dissolve iron that has settled out. —Color solution(s) is added. —Resulting pink is matched with a standard color chart, which is usually rated in ppm. —Correction is usually recommended if test shows more than .3 ppm of iron.	—Equipment used is based on amount of iron and/or manganese present and its origin. 0- .2 ppm No treatment necessary 1.0- 2.0 ppm Polyphosphate feeder .2-10.0 ppm Zeolite (ion-exchange) softener 1.0-10.0 ppm Oxidizing (manganese-zeolite) filter—usually manganese-treated green sand 3.0 and up Chlorinator and filter—usually a sand or carbon filter
Iron Bacteria —Red slime develops in toilet tanks.	Caused by living organisms (bacteria) that act on iron already in the water. Often associated with acid or other corrosive conditions.	For iron bacteria check, remove toilet tank cover and check for slippery, jelly-like coating on surfaces.	**For iron bacteria only** use: —shock chlorination of water source, pump and piping system —chlorinator and filter
BROWNISH-BLACK WATER —Fixtures stain brownish-black. —Fabrics stain black. —Coffee and tea have bitter taste.	—Manganese is present usually along with iron. —Manganese bacteria	—Test for iron in solution also measures manganese in solution, when present. Manganese bacteria cause slippery coating similar to iron except of darker color.	Same correctional methods as used for iron.
ACIDITY —"Eats away" copper and steel parts on pump, piping, tank and fixtures. —If copper or brass are being "eaten," water may leave green stains on plumbing fixtures under a dripping faucet. —If water contains iron, iron-removal methods are less effective.	Water contains carbon dioxide picked up from air, or from decaying vegetable matter, which combines with water to form a weak acid. In rare instances, water may contain mineral acid such as sulfuric, nitric or hydrochloric acids.	Test kit: —Chemical indicator is added to water sample and resulting color compared with a standard color chart to determine degree of acidity. —Measurement may be either in ppm or by "pH." If the latter, any number less than pH 7 indicates an acid condition. The smaller the number, the more acid present. If pH is less than 6, it is usually best to correct acidity.	—**Soda-ash solution**—Solution is fed into well or into suction line of pump. Can be fed along with chlorine solution or through a separate feeder unit. —**Neutralizing tank**—Limestone chips or marble are contained in a neutralizing tank in the water line. Acid is neutralized by reacting with limestone or marble. Water will increase in hardness as acid dissolves these materials.
"ROTTEN EGG" ODOR AND FLAVOR —"Eats away" iron, steel and copper parts of pumps, piping and fixtures. —If sulfur and iron are both present in water, finely-divided black particles may develop, which is commonly called "black water." Silverware turns black. —Not satisfactory for cooking	—Hydrogen-sulfide gas —Sulfate-reducing bacteria —Sulfur bacteria	—Dealers handling water-treating equipment will usually determine the amount of sulphur present from an on-the-spot sample. —Correction is needed if test shows more than 1 ppm of sulfur.	—**Up to 5 ppm**, manganese-treated green sand (oxidizing filter) can be used. —**Any quantity**, chlorinator and filter. Chlorine oxidizes sulfur, causing particles to settle out. Filter removes particles. Sulfur bacteria are killed.
OTHER OFF FLAVORS —Water may taste bitter, brackish, oily, salty, or have a chlorine odor or taste.	—Extremely high mineral content —Presence of organic matter —Excess chlorine —Water passage through areas containing salty or oily waste, etc.	Some off-taste problems are associated with the other water-quality problems listed in this table. There is no one specific test to use for other off tastes.	—**Metallic taste**—Corrected by removal of hardness or iron by methods indicated above. —**Salty taste**—Reverse-osmosis unit —**Chlorine taste and others**—Remove by passing water through an activated-carbon filter. Most other tastes and odors are removed by this method.
TURBIDITY —Water with a dirty or muddy appearance	—Silt —Sediment —Small organisms —Organic matter	Turbidity is generally measured in "turbidity units" based on the per cent of light transmitted through the water sample. If measured in parts per million, turbidity may vary from less than 1 to as much as 4,000 ppm. More than 10 ppm is considered objectionable. Health departments and manufacturers of filtering equipment make such tests since special laboratory equipment is required.	—**Fine filtering** by means of sand filter or diatomaceous-earth filter. —**Coagulation and sedimentation** by surface treatment of pond with either powdered gypsum, copper sulfate, or both. Or, use of alum in special settling tank.

with steel pipe and pump parts such as the well casing, pipe or water pump.

IRON BACTERIAL ACTION: Iron (Fe) bacterial oxidize soluble iron (Fe) to insoluble (Fe) iron, forming iron hydroxide slime to gain power to make food [they do not "feed" on iron (Fe)]. Corrosion may occur due to corrosive water OR the presence of sulfur-reducing the bacteria harbored in the slime.

You are likely to notice **dissolved iron** by the unsightly and hard-to-remove stains that develop on your plumbing fixtures. You will also see fine *rust particles* commence to form and settle, after a few minutes, at the bottom of a freshly-drawn jar of water (Figure 47a and b). This is because the iron is in solution when the water is first drawn. You are unable to see it in that form.

When water with dissolved iron is drawn from the faucet, oxygen from the air mixes with it. The **oxygen combines with the iron in solution** to form rust particles, which are easy to see. These are the particles that deposit on the inside of toilet tanks and bowls, as well as lavatories, to form a red coating. It deposits in the fabrics of laundered items causing them to become discolored (Figure 47c and d).

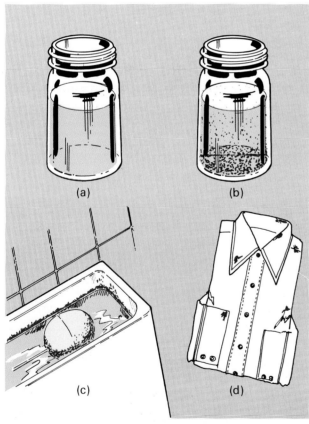

FIGURE 47. (a) Iron water, when first drawn, may be completely free of rust particles. (b) When left exposed to air, oxygen combines with the iron to form small rust particles, which settle to the bottom of the container. (c) Iron also causes red stains inside of a toilet tank. (d) Iron particles cause red stains in laundered fabrics.

Some iron particles will settle out of water when it is exposed to air in the well, or in the pressure storage tank. Most, or all, of the rust particles that develop from these exposures may settle out before they reach the faucet where you draw your jar of water. If so, the iron may accumulate and appear later in the form of slugs of red water.

Iron bacteria are an increasingly important problem that must be considered when you select a unit for iron control. You can tell if iron bacteria are present by removing the top of the toilet tank and checking for a *reddish, slippery, jelly-like substance* deposited on the sides and bottom of the tank (Figure 47c). If it is present on the tank walls, this is an indication that at least a portion of the iron content in your water supply is being caused by bacterial action.

Iron from this source will not cause particles to settle to the bottom of a jar of water.

At one time, it was generally accepted that iron bacteria "ate" the iron and steel parts of pipe and equipment because these parts become etched and corroded when iron bacteria are present. It has since been found that **iron bacteria act on the iron that is already in the water**—even as little as .1 ppm.[30] The etching and corrosion are caused by other water conditions, which happen to be the ones under which iron bacteria thrive.

It is possible to have iron present from both sources—dissolved iron and iron bacteria. If so, you will see evidence of both: particles that settle out in a water glass and slime in the toilet tank.

BROWNISH-BLACK WATER—Occasionally **manganese** is found in water along with iron, but it does not occur nearly as frequently as iron. If as much as 0.15 ppm is present, it will produce *dark brownish-black stains* on fixtures. It will settle out in much the same manner as iron except that it forms a *brownish-black sediment* and is somewhat slower in its development of particles. Fabrics washed in water containing manganese are almost certain to be stained black. For drinking water, the USEPA recommends that the water not contain more than 0.05 ppm.[26] Manganese in water is usually caused by the water passing through underground deposits of the mineral.

Some species of "iron" bacteria use manganese (Mn) in the same manner as they use iron under better conditions. The manganese residue frequently coats and clogs pipes and pump impellers and causes black staining. There are also **manganese bacteria** which appear about the same as iron bacteria. When present, they frequently cause clogging of pipes.

ACIDITY—If you found symptoms that indicate acid water, it is usually caused by the presence of **free carbon dioxide.** The carbon dioxide may come from either decaying organic matter or from rain water, which took up carbon dioxide from the air. It may be present in either surface-water supplies or ground-water supplies.

The amount of carbon dioxide present in water will usually vary from about **0 to 50 ppm.** However, there are instances where it may be as high as 300 parts per million. (Acid is also measured in pH numbers, as explained on page 53.)

In some instances, particularly in mining areas, or in areas of acid rain, free mineral acids—sulfuric, nitric or hydrochloric—may be present.

If your water supply is acid, it is important to you to correct it for two reasons:
— Acid water tends to **"eat up" metal parts** such as: the water pump, piping, tank, water heater, and fixtures.
— If you also have an iron problem, acid water **prevents complete removal of the iron.** Only a portion of the iron will oxidize.

OFF FLAVOR—If there is a rotten-egg odor or taste, note Table VIII.

If you are using **water from a pond or a lake,** it is possible that decaying organic matter or algae (small plants that form a blue, brown or red scum on a pond surface) are the cause.

If you are close to an industrial center, it may come from **chemical waste.**

If you notice an **oily flavor,** it may come from a leaky fuel tank or oil storage tank.

If you are using a chlorinator, the **chlorine taste** may increase as you increase the dosage. The taste is usually the result of the oxidation process caused by the chlorine reacting with the contents of the water. Additional chlorine, causing more complete oxidation, may actually reduce the chlorine taste. Too much chlorine may result in a free chlorine residual.

Off flavor can also be caused by extreme conditions of **hardness, iron or acidity.**

Bad taste or odor is **not usually a health hazard,** but it can make water very undesirable for drinking or cooking. Most off flavor conditions can be overcome once you find out the cause.

TURBIDITY—If you have turbid water—water with a **dirty or muddy appearance,** it is probably caused by suspended matter in the water. It is commonly caused by clay and silt that has washed in from the surrounding drainage area or from algae and organic matter that have collected in a pond or lake. Any of these cause water to have a muddy and objectionable appearance.

MEASURING THE EXTENT OF THE WATER-QUALITY PROBLEM

Table VIII, column 3, explains what **tests you can make** to measure the extent of certain of these water-quality conditions. For example, you can check water hardness, iron content (other than that caused by iron bacteria) and acidity. Making these tests requires kits which are usually available from your local water-conditioning equipment dealer. Detailed instructions are with each kit for making your own checks. However, at best, these are not too accurate. To check the other water conditions listed, you will need the help of a laboratory.

Many dealers can arrange to send a water sample to the company they represent and have a **complete analysis** made of it. This has the advantage that under laboratory conditions a much more accurate analysis can be made. Also, if you have a combination of problems, such as red water and acid water, the laboratory will determine both problems. Some health departments and educational agencies can also offer this service.

METHODS OF CORRECTING THE WATER QUALITY PROBLEM

Table VIII, column 4, gives the names of various methods for correcting water problems. These methods are explained in the next discussion—"IV. Understanding Water Conditioning Methods."

Notes

IV. Understanding Water Conditioning Methods

As a result of checking your water supply, you may have found there is need for some form of water conditioning. If you used Table VIII, you found in column 4 one or more ways of doing your water-conditioning job. This discussion will help you understand **how the water-conditioning unit or method works** that you will need. You will also learn what servicing may be needed to keep the unit(s) in satisfactory operating condition. This information will help you in selecting your water-conditioning unit. It will enable you to appreciate the amount of servicing needed and the importance of proper maintenance.

When you are ready to actually select the water-conditioning system—that is, to determine size, type, etc.—you will need help from a dealer. He should understand your local conditions and be in position to install and service your unit(s) when you have trouble. In some areas you can rent water-conditioning equipment. This is especially helpful if there is some question as to how satisfactory a particular unit may be.

There are *seldom more than one or two water-conditioning problems with any one water source.* Consequently, some of this information will not apply to your situation.

Research is being conducted on all types of water conditioning equipment or sponsored by EPA. For example:
1. Granular Activated Charcoal GAC Filters.
   —Portable, pour-three models.
   —Faucet models.
   —Stationary models.
   —Lime by-pass models.
2. Reverse osmosis units.
3. Distillers.
4. Ozonators.
5. Others.

Select the water-conditioning situation found in your water supply from the following headings:

A. Means for Controlling Water Hardness.
B. Means for Controlling Iron in Water.
C. Means for Controlling Manganese in Water.
D. Means for Controlling Acid in Water.
E. Means for Controlling Off Flavor in Water.
F. Means for Controlling Turbidity in Water.

A. Means for Controlling Water Hardness

There are many areas of the country where water hardness is not a problem, or at least not a serious one. Figure 48 shows the variations in water hardness by areas throughout the country. Areas with water of less than 3 to 3.5 grains of hardness per gallon (slightly hard) have no problem.

Various chemicals are available for adding to water to soften it. But, the most satisfactory method for home use is the type of softener that connects into the supply line of your pumping system. Those are the types explained in the discussion that follows:

They are:
—Reverse-osmosis units.
—Ion-exchange units.

REVERSE-OSMOSIS UNITS

Reverse-osmosis units are relatively new. They are used to desalt (Na, Cl, SO_4) individual water supplies as well as remove hardness. They can be secured in almost any size desired, but the cost is much greater than for ion-exchange units, which are discussed under the next heading. For drinking water, they have an advantage over ion-exchange units. Sodium, which ion-exchange units add to water in the exchange process, is not recommended for people who have heart, kidney, or circulatory ailments.[21] The reverse-osmosis units add nothing to the water they process.

Before you can understand what is meant by the term "reverse osmosis," you need to know what is meant by the term "osmosis." You may remember it from experiments in your science class in which your instructor used a special membrane to separate salt water from fresh water. He showed that fresh water will move through the membrane to the salt-water side (Figure 49a). With fresh-water and salt-water levels the same at the start, the salt-water level will gradually rise and the fresh-water level lower as the fresh water moves through the membrane to the salt-water side. This action is called "osmosis." This is an action that takes place in nature, especially in plant growth.

The important point in this illustration is that the **fresh water** was able to move through the membrane, while water containing salt did not move through the membrane to the fresh-water side.

This straining action (called "selective permeability") is important in **reverse osmosis,** as shown in Figure 49b. In this case, water containing calcium, magnesium, and even small dirt particles, bacteria, and virus can be pumped under pressure into the chamber on the left side of the membrane. Under pressure, water

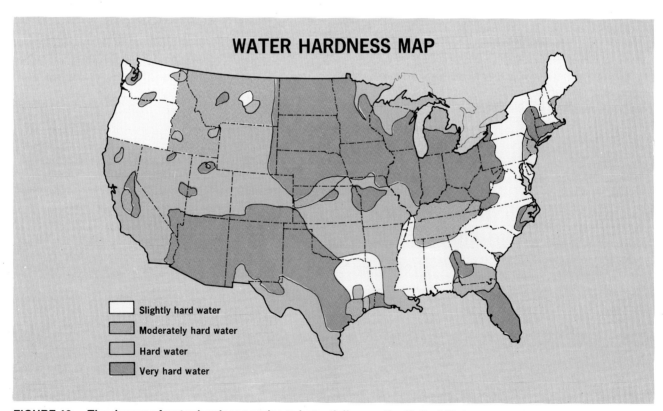

FIGURE 48. The degree of water hardness varies substantially over the United States.

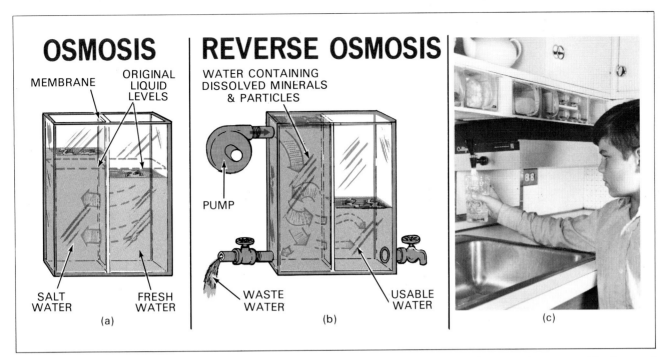

FIGURE 49. (a) Osmosis involves movement of fresh water through a membrane into the side containing solubles (saltwater). (b) Reverse osmosis involves applying pressure to the water on the side containing salt, hardness particles, etc. Water forced to move through the membrane in reverse to its normal movement is cleansed of the soluble materials it contains. (c) Commercial reverse-osmosis drinking-water unit. These are sometimes called purifiers, because they also remove some bacteria and viruses as well as hardness particles.

passes through the membrane in the reverse direction from normal osmosis, thus, the term "reverse osmosis."

This action leaves almost all of the hardness particles and much of the bacteria and virus on the pressure side of the membrane. Enough of the *hardness particles are strained* out so the water is considered soft. As the hardness particles build up in the pressure chamber, it becomes necessary to flush them away through a waste outlet. The unit is then ready for continued operation.

Since the reverse-osmosis **process is relatively slow,** it is necessary to operate the unit almost full-time and to store the conditioned water for those times when there are heavy demands.

These units are commonly fitted with an **activated-carbon filter,** which helps remove some of the larger particles from the incoming water and, at the same time, improves the taste (Figure 49c).

Most of the under-the-sink type units produce about 5 gallons of water per day. Larger RO units are also available to treat the entire water supply if necessary.

The reverse-osmosis softener requires very little **servicing.** The membrane will need to be changed about every one to two years as it becomes fouled with hardness particles. If the unit is equipped with an activated-carbon filter, the filter unit will probably need to be cleaned or replaced about twice a year.

ION-EXCHANGE UNITS

An **ion-exchange water-softening unit** consists of a tank containing sand-like water-softening material known by such names as "zeolite," "exchange material," "minerals," and "ion-exchange resins."

The zeolite will not dissolve in water, but when hard water is passed through it, **sodium ions from the zeolite are exchanged for the calcium and magnesium (hardness) ions** of the water. For this reason, zeolite is known as an "exchange material." The sodium ions added to the water do not cause hardness. Figure 50 shows an animated concept of the ion-exchange principle.

If you **do not understand about ions** or have forgotten, maybe this explanation will help. Ions are electrically-charged particles that develop when certain substances are dissolved in water. Common salt (sodium chloride) is one of them. In solution, the sodium-chloride molecule separates into sodium ions (which are positively charged, Na+) and chloride ions (which are negatively charged, Cl−). Water tends to insulate these particles from each other so they will not recombine.

The **molecules that cause hardness** —such as calcium and magnesium carbonates—**ionize in water in the same manner.** It is in this ionized condition that the exchange takes place in the zeolite exchange material. The sodium ions in the exchange material trade

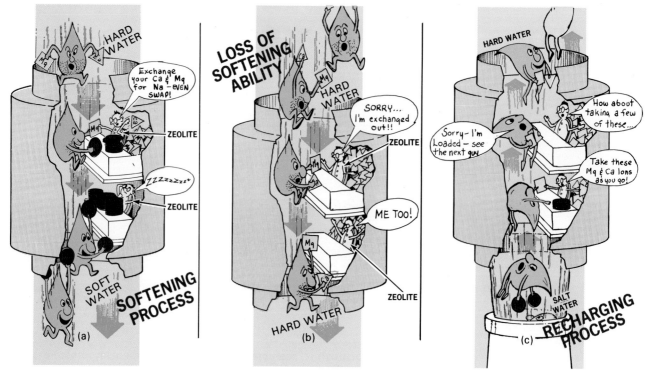

FIGURE 50. Zeolite ion-exchange principle of water softening. (a) As hard water, containing calcium (Ca) and magnesium (Mg), passes through the zeolite, the Ca and Mg ions are exchanged for sodium (Na) ions which do not cause water hardness. (b) As the zeolite collects hardness ions, it finally uses up its capacity to soften until it is recharged. (c) To recharge, salt solution is flushed through the zeolite. Sodium (Na) ions from the salt replace the Ca and Mg ions in the exchange material. The flush water containing the hardness ions is drained out of the system in the opposite direction and wasted.

places with the "hardness" ions of calcium and magnesium in the water.

After the **zeolite is exhausted** —has exchanged its supply of sodium for calcium and magnesium—any additional water that passes through continues to be hard.

To **recharge the exchange material** and make it useful again, all you need to do is pass a concentrated salt brine solution through it. The zeolite then releases the calcium and magnesium ions and exchanges them for the sodium ions from the salt. This recharges the zeolite. Excess salt is then washed out of the exchange material. The unit is ready to soften more water. About ½ pound of salt is needed to supply enough sodium to replace 1,000 grains of water hardness. The zeolite lasts indefinitely if recharged regularly.

If the home plumbing systems are not arranged to soften only the bathtub, lavatory, kitchen sink and laundry, another alternative is to soften only the hot water (partial softening) line. All home systems are arranged to do this rather easily.

Softened water is generally *supplied only to the bathtub, lavatory, kitchen sink, and laundry.* Supply lines to toilets and outlets for general use are supplied with unsoftened water. This reduces the total water passing through the softener unit, thus reducing the frequency of recharging.

According to health authorities,[21] **some persons with certain conditions should not consume water containing sodium** such as that from an exchange water softener:

The presence of sodium in water may affect persons suffering from heart, kidney, or circulatory ailments. Persons so affected should rely on their physicians' advice. When a strict sodium-free diet is recommended, any water should be regarded with suspicion. In light of the preceding facts and because individual intake of sodium varies, no recommended limit for sodium has been established.

The **maintenance required** with an ion-exchange softener depends largely on the ease with which it can be recharged. This, in turn, depends on the design of the softener. Softeners are of two general designs:

— **One-tank units**—consisting of the mineral (zeolite) tank only (Figure 51a).
— **Two-tank units**—consisting of a mineral tank and a brine storage tank (Figure 51b).

To **recharge the one-tank unit,** salt is emptied directly into the top of the tank where it mixes with water in the tank to form brine.

With the **two-tank unit,** brine solution is prepared in a second tank. When needed for recharging, the brine is siphoned into the zeolite tank from the brine tank. Ready-prepared brine solution is more efficient for

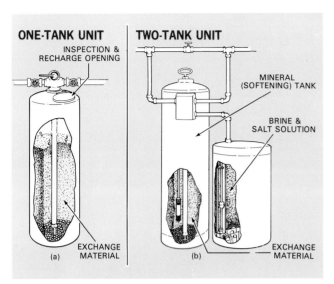

FIGURE 51. Types of water softeners. (a) One-tank unit. Salt for recharging is added in dry form into top of mineral tank. (b) Two-tank unit. A salt-brine supply is kept in a separate tank and siphoned into the mineral tank when needed.

recharging the zeolite than dry salt emptied into the one-tank units.

With either design, during the period **when the mineral tank is being recharged,** provision is made, in the plumbing or softener-control valve, to bypass hard water into that portion of the plumbing system served by the softener. During that time, only hard water is available for all purposes. For that reason, users of *non-automatic softeners* often recharge the exchange tank just before retiring at night. In the morning it is ready for draining, flushing, and reconnecting to the plumbing system.

The **frequency of recharging** varies with the size of the unit, the hardness of the water, and the amount of water used. Some require charging daily. Others may not need recharging for two or three weeks.

Most softeners being sold at the present time are of the two-tank type and are **automatically recharged.** All steps in the recharging operation are handled automatically by means of an electric motor which works the automatic controls. Recharging is simply a matter of tripping a switch at the start of the operation. Units are also available where even the switch is controlled by an automatic timing mechanism or by a sensor (Figure 52a) that detects when the conditioned water is starting to get hard.

The **salt used in water softeners** is specially prepared for that purpose. It is more efficient because it is free of dirt and free of insoluble material that tends to foul the zeolite bed.

If the **exchange material is not recharged regularly,** slime and dirt may collect in the exchange bed. This will cause the exchange material to gradually lose its capacity to soften. You should avoid letting the zeolite get in that condition. But, if it does happen, there are chemicals available which you can use to activate the exchange material, or the old zeolite can be replaced with new zeolite.

If your **water supply contains dissolved iron** and it is exposed to oxygen, small particles of the oxidized iron may accumulate in the zeolite bed. In that case you will need to backwash the zeolite occasionally to keep the iron particles from accumulating and fouling the bed. You can also get special salt, or chemicals to add to the regular salt, which will help prevent fouling of the zeolite bed. Filtration is usually recommended.

In some areas maintenance is provided by a **water-softening service.** It is available for a monthly charge. A service company owns the equipment. A freshly-charged, ion-exchange softener tank is delivered to your home to replace the one you have been using (Figure 52b). The used one is then recharged at a central plant and put into use again.

Another type of service consists of exchanging the **zeolite bags.** To recharge the softener, the exhausted bags of zeolite are removed from the tank and replaced with recharged bags of zeolite. The exhausted zeolite is then returned to the service company for recharging.

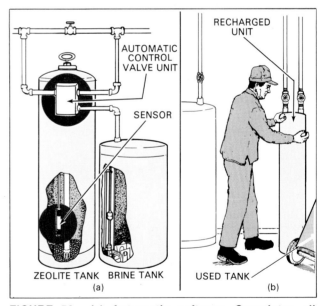

FIGURE 52. (a) Automatic softener. Completes all recharging steps automatically. (b) Replaceable unit supplied by a service company. Used tanks are replaced with recharged tanks at regular intervals.

B. Means for Controlling Iron in Water

If you found that iron is present in your water supply, and if it contains more than 0.3 ppm of iron, it is not considered satisfactory. Most water containing iron ranges below 5 ppm with some ranging as high as 15 ppm. An extreme condition may be as much as 60 ppm.

As explained earlier, if **dissolved iron** is present in your water supply, it will appear as red particles that settle out when exposed to air or as red slime if **iron bacteria** are present. It is also possible to have both red particles and slime. To make the situation more difficult, some methods of treatment work only with dissolved iron. Only one method of treatment works for both—control of iron bacteria and dissolved iron.

The American Water Well Association prefers in-well treatment to protect pumps, pipe and casing.

The most common methods used to meet the varying iron conditions in farm and home water supplies are the following:
— Phosphate feeders (dissolved iron)
— Ion-exchange units (dissolved iron)
— Oxidizing filters (dissolved iron)
— Chlorinator-and-filter units (dissolved iron or bacterial iron, or both)

The method of iron control you select will be influenced by: (a) the *amount of iron present* in ppm (parts per million), (b) *what form it is in,* and (c) what *other water quality problems* you may have that can be handled by the same water-conditioning unit.

How each of these methods work is explained in the following discussion.

PHOSPHATE FEEDERS

Phosphate feeders will handle up to 2 ppm of dissolved iron. These units are *not satisfactory if the iron is oxidized and already settled out as rust particles. It will also be unsatisfactory if the iron condition is caused by bacteria.*

These units use a **food-grade phosphate** known by such commercial names as "Zeotone," "Micromet," and "Nalco M-1." Exactly how the phosphate works is not known, but it is believed that it coats the iron in solution so, when it is exposed to oxygen, there is no oxidizing action. However, if the treated water is boiled over a period of time, some iron may settle out. The phosphate treatment keeps the water clear and satisfactory for drinking purposes. It also protects iron pipe from mineral deposits.

When in operation, the polyphosphate compounds

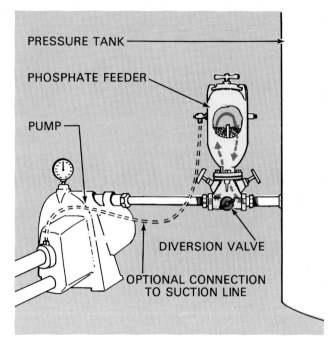

FIGURE 53. Phosphate feeder installation. As shown, a portion of the water being delivered by the pump is directed through the feeder where it dissolves some of the phosphate material and returns to the water line. Some feeders are connected so the phosphate water returns to the suction line. This is desirable for pumps that supply air for the pressure tank along with the water.

are fed through a unit of the type shown in Figure 53. You can also use a **chemical feeder** similar to those used for adding chlorine to water (Figure 29).

In Figure 53, note that the feeder is connected into the piping ahead of the pressure tank. The reason: iron must still be in solution for the phosphate to be effective. Once the iron-laden water has been exposed to air in the pressure tank, rust particles start to form, and the phosphate material is not effective in controlling it. The iron will then continue to produce stains.

Phosphate-feeder units are often recommended for **installation with a water softener.** They keep iron particles from developing, then settling in the zeolite and fouling it. When iron particles accumulate in zeolite, they plug the pores and gradually reduce softening capacity.

Maintenance involves keeping the feeder supplied with phosphate material. About one pound of phosphate material is needed for each 60,000 gallons of water containing 2 ppm of iron. Consequently, the maintenance required will be directly related to the size of your phosphate feeder and the amount of iron in your water supply, as well as the amount of water you use. Feeders are commonly sized to hold about a 30-day supply of phosphate.

ION-EXCHANGE UNITS

An ion-exchange zeolite unit is for water that has not been exposed to air thus causing iron particles to form. Under these conditions ion-exchange units can be effective for water containing up to 10 ppm of dissolved iron.

There is a **wide selection of zeolite minerals,** some of which will remove iron on an ion-exchange basis the same as described for removing calcium and magnesium when softening water. Or, they will remove iron along with calcium and magnesium as part of the softening process. If your water supply has been checked for iron content, a manufacturer of water-conditioning units can take the information and select the zeolite needed for your water supply. The unit will look like a water-softening unit, so there is no way of telling from the outside appearance whether a softener will handle small quantities of iron or not.

If your water supply has been exposed to air and **already contains iron particles,** they will be filtered out on top of the zeolite bed as the water moves through it. They are quite difficult to remove. The reason—the iron particles are heavier than the zeolite. When you attempt to backwash the zeolite bed with a heavy flow of water, the zeolite particles move more readily than the iron particles.

If you leave the iron particles in the zeolite bed, they will accumulate until they bleed through the bed. Also, the bed will become fouled until it loses its ion-exchange capacity. Once the zeolite bed reaches this condition, it is possible to get a pellet-type salt containing a substance which dissolves the iron and rust accumulations.

If you have a situation where part of the iron is forming into particles, use an oxidizing filter (next discussion). It is designed for both dissolved iron and iron particles.

If **iron bacteria are present,** a zeolite softener is not satisfactory. Slime that develops with the bacterial action gradually clogs the zeolite, causing it to become less and less effective.

The ion-exchange units are **serviced** in the same way and with about the same frequency as described for the water-softener ion-exchange units.

OXIDIZING FILTERS

Iron-removal (oxidizing) filters are used for removing up to **10 ppm** of iron unless iron bacteria are present. The slime that develops as a result of bacterial action will clog the mineral bed until its iron-removal action will not be satisfactory.

The iron-removal-filter unit looks much like a water softener, but the material contained inside is some type of oxidizing material, usually **manganese-treated green sand.** The manganese provides oxygen which causes the iron to settle out as rust particles. If some iron particles have formed before the water reaches the filter, they are also filtered out by the mineral bed.

Improved iron filters that do not require green sand are now available.

Maintenance consists of backwashing and rinsing the filter about *every week* and recharging, either then, or at longer intervals. Most of the iron particles settle in the lower half of the mineral bed. It is highly important that the bed be backwashed regularly to keep this accumulation of rust particles from passing out as "slugs" of iron particles during a period of heavy usage.

Backwashing consists of passing water through the filter in the opposite direction from the regular flow. This action removes the accumulated iron particles. Considerable water is needed if the backwash is to remove the iron particles satisfactorily. Flow rates as high as 8 to 10 gallons per minute per square foot of area in the mineral bed are often recommended. You will need to take this into account in selecting a pump with enough capacity to supply this need.

The iron-removal (oxidizing) filter is **recharged with potassium permanganate.** The potassium permanganate is placed in the top of the tank, and the unit slowly rinsed with a down-flow of water. This chemical recharges the mineral with oxygen, in a similar manner to the way salt recharges a softener. Recharging may vary from **weekly to monthly,** depending on the size of your filter and the amount of iron in your water supply.

An aeration-filtration system is available that requires no chemicals. It is a single-pumping system and operates under closed well pump pressure. It is called the Ferr-X system. This unique system operates under compressed air and the water is never exposed to outside contamination.

CHLORINATOR-AND-FILTER UNITS

A **chlorinator and filter** will take care of **moderate quantities of iron from any source**. The chlorine oxidizes the iron so that rust particles form readily. A filter strains out the iron particles and provides clear water at the faucet. Figure 54 shows a combination chlorinator and filter.

Where **iron bacteria** are the cause of iron in the water supply, chlorine will kill them and protect the system from further attack.

This unit may be overwhelmed by severe infestation, iron bacteria.

For a chlorine unit to be most effective, it is important that the **water contain little or no acid**. This is another point your dealer will take into account in recommending a system for your needs. If considerable acid is present, it is easily corrected by mixing soda ash or caustic soda with the chlorine solution. The mixture is then fed through the chlorinator unit. This is discussed further in the section, "Means for Controlling Acid in Water."

Maintenance of the chlorination unit for iron control is the same as discussed on page 47 for disinfecting purposes. For maintenance of the filter that is used with it, see page 66.

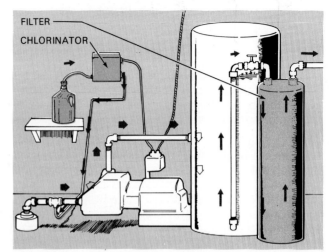

FIGURE 54. A combination chlorinator and filter will remove moderate quantities of iron in the water supply. A standard chlorinator unit meters chlorine into the suction line to provide the oxygen that causes the small particles of iron to form. The iron particles are then filtered from the water by the filtering unit next to the pressure tank. If the iron is the result of bacteria, chlorine kills the bacteria before they are distributed through the supply line to form slime and discoloration on the plumbing fixtures.

C. Means for Controlling Manganese in Water

If manganese is present in your water supply, you are almost certain to have iron present also. If as much as .15 ppm is present, you will need to provide some means of control. Fortunately, the *same control methods are used for dissolved manganese as are used for dissolved iron*. Likewise, the same control is used for manganese bacteria as for iron bacteria.

D. Means for Controlling Acid in Water

If the check you made of your water supply indicated an acid condition that needs to be corrected, there are two units you can use. They are:
— Soda-ash or caustic-soda feeder
— Neutralizing tank

Here is how each of these units work.

SODA-ASH OR CAUSTIC-SODA FEEDER

Soda-ash is particularly well adapted for the problem where you are already using a chlorinator. The soda-ash and chlorine solution can be mixed together and fed through the same chlorinator unit (Figure 55a).

If you do not have a chlorinator, a **chemical feeder** of the same type is satisfactory for feeding soda-ash solution by itself.

The best practice is to feed the soda-ash solution into the well. You then provide protection against acid water corroding the well casing, the pump and the piping system.

Soda-ash **adds no hardness** to the water. It is a relatively harmless material that has no marked effect on water used for bathing, drinking or clothes washing. However, it does add sodium bicarbonate to the water.

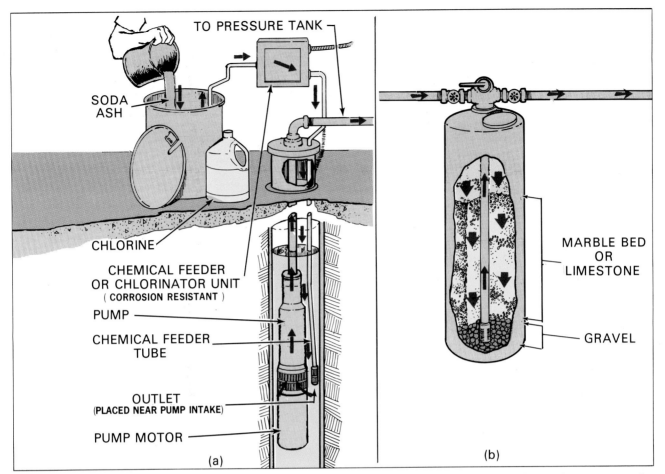

FIGURE 55. Methods used to neutralize acid water. (a) Addition of soda-ash or caustic soda to the water supply by means of a chemical feeder. If the chemical feeder is already being used to add chlorine to the water supply, soda-ash or caustic soda can be added to the same solution. (b) Use of a neutralizer tank.

If this is objectionable, you can use caustic soda instead of soda ash.

Maintenance consists mostly of making up the soda-ash solution. About ½ pound of soda-ash is added to one gallon of soft water. It is then fed through a feeder in the same manner as chlorine solution, or it may be fed with chlorine solution. Soda-ash solution will need to be replenished about every two to three weeks. Soda-ash is available in 100-pound bags from chemical-supply houses. Soda-ash adds no hardness, but does add sodium.

Caustic soda is the same as household lye. It is stronger than soda-ash and requires more care in handling than soda-ash. It, too, is available from chemical-supply houses. If you use it, follow the directions given by the manufacturer when mixing the solution.

NEUTRALIZING TANK

The neutralizing tank is similar in appearance to a water-softener tank except it contains a bed of **limestone or marble chips** (Figure 55b). The acid reacts with these materials and gradually "eats" them until they have to be replaced. This action neutralizes the water until most of the corrosive action has been overcome.

The **flow rate through the neutralizer is slow** because some time is needed for the acid to react with the limestone. It is sometimes necessary to use two neutralizers hooked in parallel to provide enough neutralizing capacity to match the water flow rate.

Since a small amount of limestone is dissolved in the water to neutralize the acid, **water hardness will increase** slightly. The increase in hardness can be overcome by installing a water softener immediately after the neutralizer.

Maintenance of a neutralizer requires *backwashing* regularly—perhaps weekly—to loosen and clean the neutralizing bed in the tank. Your dealer will tell you how frequently. The neutralizing material is heavy, so fairly high backwash rates are required to loosen and clean the bed.

Since the limestone neutralizing bed is dissolved with use, it must be checked each year, and the *dissolved portion replaced*.

E. Means for Controlling Off-Flavor in Water

As shown in Table VIII, taste and odor conditions can develop from a wide variety of causes. Many of them can be overcome with one of the units described in the foregoing discussion. These include **rotten-egg odor and taste, metallic taste, salty taste, and chlorine taste.**

If your water supply has a **bitter, fishy, marshy, brackish or oily taste,** you can usually overcome the problem with an activated-carbon filter.

If you are using **pond water,** the blue-green scum that forms on the water may be the cause of an off taste or odor. The scum consists of plants called "algae." They also cause turbidity. Consequently, algae control is discussed under the next heading "Means for Controlling Turbidity in Water."

Check with your U.S. Environmental Protection Agency office for the latest research on these types of filters.

The following discussion is to help you understand about activated-carbon filters.

ACTIVATED-CARBON FILTERS

There are two types of activated-carbon filters: (1) *the pre-coat cartridge type* (Figure 56a) and (2) the *carbon-bed type* (Figure 56b).

The activated carbon used with both filters is the means by which taste and odors are removed.

Activated carbon is made from bituminous coal, lignite, petroleum coke, and peat. After preliminary processing, these materials are heated and reacted with steam to develop the extensive internal pore structure required for absorption. This is called **"activation."** Activated carbon can act as a filter to remove finely divided solids, but its primary function is to remove organic compounds that cause tastes and odors, or organic compounds that are potentially harmful, such as pesticides or cleaning solvents that occasionally are detected in drinking water sources. Activated carbon is also used to dechlorinate water.

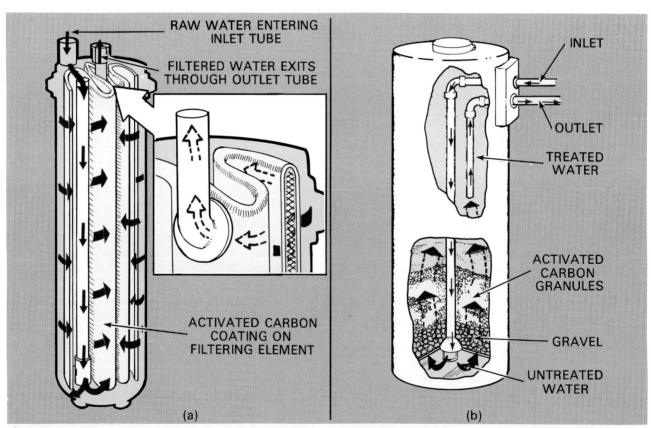

FIGURE 56. Types of activated carbon filters. (a) Cartridge-type pre-coat filter consisting of an activated-carbon coating on a filter element. Water is filtered as it passes from the outside of the filter element into the inner area. (Inset) Filtered water inside the element is collected and discharged through a special outlet. (b) The bed-type unit consists of a tank with a bed of activated carbon granules in the bottom.

Maintenance of the *cartridge-type, pre-coat filter* of the smaller sizes consists of replacing the entire cartridge when the water pressure starts to lower noticeably. The slow water flow is caused by particles building up on the surface of the filter and slowing down water movement.

With the *larger types of pre-coat filters,* the filter is removed and cleaned by washing off the dirt particles that have accumulated. After several cleanings, and when the carbon material has absorbed all the taste and odors it can hold, the filter is replaced.

Normally, a pre-coat filter can remain in operation over a period of several months before it needs to be changed or washed.

The maintenance required for a *bed-type, activated-carbon filter* consists of backwashing the filter bed to remove suspended dirt from the filter bed. Under most conditions, you will not need to backwash more than once or twice a month.

The carbon bed gradually **absorbs tastes and odors until it becomes saturated.** Then it must be replaced. For an average household, the activated carbon will last from about one to three years.

Activated charcoal systems come in various sizes. The life of the carbon system is also very dependent on the amount of carbon in addition to the intensity of the taste and odor and the quantity of water treated.

Some home systems are also available to remove excessive amounts of fluoride, nitrate, arsenic and other chemicals. The National Sanitation Foundation has recently established a testing program to evaluate these types of systems.

F. Means for Controlling Turbidity in Water

If you are troubled with turbidity, your source of water is probably a pond, lake, or stream. **"Turbidity" is suspended matter** which collects in water—such as **clay, silt, algae, and organic material**—which either washes in from the drainage area or develops in stagnant water. It causes water to have a muddy and objectionable appearance.

Turbidity, if measured in parts per million, varies from less than 1 to as much as 4,000 ppm by weight.

The ppm method of measurement has been largely replaced by another called "turbidity units." They are determined by a system which measures the scattered light from a sample of turbid water. For measuring turbidity of pond water, the two methods of measurement are about equal (see table on page 71).

If you need to determine the **amount of turbidity** in your water supply, you can have a sample checked. Manufacturers of filters, or your local health department, may be able to measure it. If the turbidity is greater than 10 ppm, the water is usually objectionable and will not meet public health standards in many states.

Considerable research has been done on **methods of removing particles** that cause turbidity in an effort to find the most practical solution. There is still some difference of opinion about the most effective means of treatment. However, there is general agreement that water, drawn through such *makeshift filters* as a barrel filled with stones and sand and placed in the bottom of the pond or a sand-filled trench connecting the pond with a storage reservoir, has only the larger particles filtered out.

Other filters, which are sometimes considered, include those that use **pads, ceramic cylinders, paper and porous stone.** These are not considered satisfactory for filtering turbid water because of low capacity, the ease with which water can channel, their tendencies to crack, or their inability to filter out fine particles.

There are three approaches to controlling turbidity: (1) the open-pond method, (2) the use of turbidity treatment units, (3) a combination of both methods. How these different methods work is discussed under the following headings:

Open-pond treatment:
— Coagulation and sedimentation
— Copper-sulfate treatment (algae)

Turbidity treatment systems:
— Sedimentation and filtering system
— Diatomite filter
— Rapid sand filter

Seldom is it possible to use open pond treatment without also using one of the turbidity treatment systems if you want clear water. The purpose of the open pond treatment is to take a heavy part of the turbidity load off of the treatment system at those times when turbidity is greatest. For example, the amount of suspended matter—dirt, silt, and fine sand—may be very sizeable after a heavy rain. Then, during certain seasons of the year, the blue-green scum on the surface of the water (algae) can cause high turbidity unless controlled. Using the open pond treatment to control these conditions will greatly reduce the load on your turbidity treatment system.

Open-Pond Treatment

COAGULATION AND SEDIMENTATION

Coagulation and sedimentation is a process that causes the fine sediment in the water to collect into larger particles and settle to the bottom of a pond before the water reaches the filter. This can be accomplished with **powdered gypsum.** It can be spread by hand over the surface of the pond at the rate of about 12 pounds per 7,000 gallons of estimated water storage in the pond (Figure 57). The powdered gypsum causes the turbidity particles to collect in clusters (coagulate) and settle to the bottom (sedimentation).

Powdered gypsum is safe for both marine life and humans if applied in the quantity recommended for this purpose.

COPPER-SULFATE TREATMENT (ALGAE)

Algae—the blue-green or reddish scum that forms on the water surface during summer—contributes to turbidity and to an offensive odor and taste. To control it, **copper sulfate** can be added to the water to kill the

FIGURE 57. Gypsum may be applied to pond surface to clear up turbidity without killing fish or other marine life.

algae. Apply at the rate recommended by your local health department.

Copper sulfate is safe for marine life and humans if applied in the quantities recommended.

Turbidity Treatment Systems

A turbidity treatment system may be made up of **one or more units** that are required to meet the turbidity problem. They have to be sized to meet your water requirement needs, and selected on the basis of the amount of turbidity and the size of particles to be filtered. When you are ready to install one, you will need help from your local health department, your dealer or your state agricultural college.

This discussion is to help you understand how each of the systems work, their limitations and the amount of maintenance required by each one.

SEDIMENTATION AND FILTERING SYSTEM

If **turbidity is a continuing problem** and amounts to **more than 40 units,** you will need a turbidity treatment system. The system commonly used for pond or lake water consists of (a) a screened intake and piping for conducting the water from the pond to the first unit, (b) a settling tank for coagulation and sedimentation of the turbidity particles, (c) a filter, and (d) a storage compartment (Figure 58).

If **turbidity is less than 40 units,** you may be able to cut out the settling tank and let the filter do the whole job.

In Ohio studies[29] the turbidity of pond water ranged from 1 to 165 units. The lowest average turbidity was at a depth one foot below the water surface. At that level turbidity was about 21 units. Consequently, most recommendations are that the **screened intake** be located about 12 to 18 inches below the pond surface (Figure 58) to take advantage of less turbid water.

If a **settling tank** is used, the pond water enters it first. Figure 58 shows such an arrangement where a settling tank is used with an alum (aluminum sulfate) feeder.

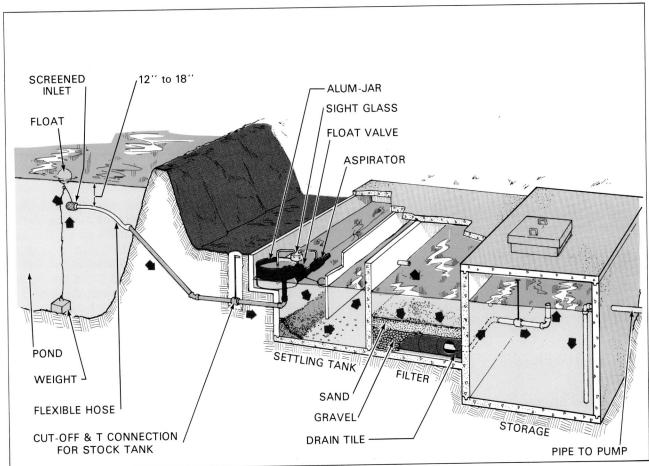

FIGURE 58. A system for removing turbidity from water. Screened inlet in pond is positioned to take water from the cleanest level in the pond. Water enters through the screened inlet and is conducted to the settling tank (coagulation and sedimentation chamber) first. Alum (aluminum sulfate) is added. After a major portion of the suspended particles are settled out, the water flows into the sand filter where essentially all of the remaining particles are removed. After passing through the sand filter, the clear water enters the storage compartment.

The **alum solution** is used to speed up sedimentation. It is fed into the settling chamber in proportion to the amount of water that is feeding in from the pond. The alum forms a "floc," or gelatin-like glob. The small particles collect and stick to the floc which gradually settles to the bottom of the settling tank. Now there are also polymer compounds available which are more effective than alum.

Maintenance consists of cleaning the settling tank occasionally to remove the mud and floc accumulation in the bottom of the tank. With a well-designed system you would probably not have to do this more than about every 6 to 8 months.

If you are using an **alum feeder,** you may need to add alum solution about every two weeks. You will need experience to determine the frequency. If the incoming water is fairly clear, much less alum solution is needed than when turbidity is high.

If a settling tank is used, the next unit is most likely to be a **slow sand filter.** If a settling tank is not used, the pond water will enter the filter directly.

The slow sand filter is the type used most extensively for pond-water treatment. It is relatively low cost and very effective if properly maintained. As shown in Figure 58, it consists of a second compartment into which the water moves after leaving the settling tank. The filter usually has a layer of coarse gravel on the bottom and, on top of that, about 18 to 24 inches of fine sand.

Water moves through the filter by gravity. This means that it moves fairly slowly—about 27 to 75 gallons per day per square foot of filter-bed surface. Consequently, it is necessary to have a rather large surface for the water to filter through to provide enough water for daily home and farm needs.

As the **water deposits the dirt particles,** they build up on top of the sand layer and assist in the filtering action. In fact, the filter becomes more effective in removing dirt particles as the dirt layer thickens. However, the amount of water being filtered gradually decreases until the time is reached that the dirt layer must be removed.

Maintenance consists of removing the dirt and about ½ to 1 inch of top sand, when the filtration rate becomes too slow. The time for this may vary from two weeks to six months. Frequency of cleaning depends on the amount of water being filtered, the amount of solids it contains, and the size of the filter bed. After about 6 inches of sand has been removed from the top of the filter bed, it should be replaced with clean sand.

From the filter the water is drained into a **storage tank or reservoir** (Figure 58). The storage tank should be large enough to take care of peak demands so that you are not dependent on the rate of water flow through the filter to take care of your immediate needs.

Maintenance amounts to checking the storage compartment occasionally to see that no dirt is entering from the surface, that the side walls and floor are not cracking, and that there is no accumulation of sediment in the bottom of the compartment. If the filter is doing a satisfactory job of removing dirt particles, the storage compartment may not need cleaning for several years.

DIATOMITE FILTER

Diatomite filters—sometimes called "diatomaceous-earth filters"—are now being used effectively for removing turbidity from water.

FIGURE 59. One type of diatomaceous-earth filter. When filter is cleaned, new filter material (diatomaceous earth) may be applied to the filter fabric by hand or fed into the incoming water supply. With the latter method, a filter cake builds up on the fabric.

The difficult term "diatomaceous earth" describes the material used to provide the filtering action. The material comes from the remains of a marine algae called "diatoms." These are extremely small plants that develop shell structures. When the plants die, the shells accumulate to form diatomaceous earth. It is estimated that the shells of as many as 50 million diatoms are present in a cubic inch of material. This large surface area provides excellent filtering action.

Studies conducted at Purdue University[32] showed that no **coagulation and sedimentation** is needed for this type of filter.

Figure 59 shows one of these units. The **filtering element** usually consists of a porous surface called a "septum." It may consist of wire cloth, plastic fiber cloth, or any of several materials that will let water pass readily. A coating of **diatomite-filter material** is then applied to the septum to form a "precoat." It is this material that provides the filtering action when the filter is first put into operation.

A diatomite filter removes the suspended solids by **straining action,** the same as a sand filter, but it is much more effective. It filters 10 to 100 times faster than a slow sand filter and removes a larger amount of the very small particles.

The filter tank is installed in the plumbing system, on either the **delivery side of the pump or on the suction side.** The latter is often considered the best arrangement since the tank can be more easily opened for inspection and cleaning at any time.

Maintenance of diatomite filters consists of adding diatomite-filter-aid material as the filtering action starts to slow. This increases the flow rate through the filter. The point is finally reached where further addition of filter aid has little effect. At this stage the entire filter cake must be removed and new diatomite-filter material added to the septum to form a new cake. If your filter is properly sized to your water use, the filter cake should not require replacing more than about every 2 months.

If a diatomaceous earth filter removes large amounts of biological particulate matter, such as algae, this may decay and cause taste and odor problems. If this occurs, the filter cake should be replaced more frequently before the algae has time to decay.

RAPID SAND FILTER

The rapid sand filter—sometimes called a "pressure filter"—is a **tank-type arrangement,** similar in appearance to the neutralizer tank shown in Figure 55b. It contains fine sand which rests on top of a bed of coarse sand and gravel.

A rapid sand filter is usually **connected into the delivery line** from the pump so it works under whatever

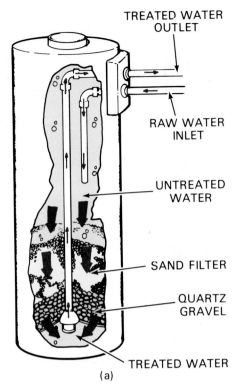 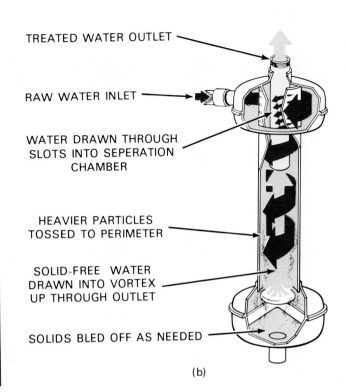

FIGURE 60. Rapid sand filters. (a) Filter type (b) centrifugal separator.

MAXIMUM LEVELS OF TURBIDITY, COLOR AND SECONDARY CONTAMINANTS ACCEPTABLE FOR DRINKING WATER	
(U.S. Environmental Protection Agency)	
Aesthetic Quality:	(Level)
Turbidity	1 NTU[a]
Color	15 PCU[b]
Contaminant:	
Chloride	250.
Copper	1.0
Fluoride	1.4 to 2.4[c]
Iron	0.3
Manganese	0.05
Nitrate, as N	10.
pH	6.5 to 8.5[d]
Sulfate	250.
Zinc	5.0
Total Dissolved Solids	
TDS	500.
Sodium	160[d]

[a]Nephelometric turbidity units
[b]Platinum cobalt units
[c]Where average annual maximum daily temperature ranges from 90.5° to 53.7° F and below
[d]No units applicable
[e]Taste threshold—not firmly agreed upon

pressure is developed by the pump. Consequently, its filtering capacity should be in relation to the capacity of the pump. This is usually figured around two to three gallons per minute per square foot of sand surface. Water flows through the sand bed from top to bottom. The dirt particles collect on top of the filter bed.

A pressure filter works well if the **particles being filtered are large and limited in quantity.** However, it is the least effective of the three filters for small-size particles such as clay particles. Because of this characteristic, a rapid sand filter is not usually considered satisfactory for use with pond water unless supplied with a filter aid.

Filter aid is a chemical solution which is added to the water with a liquid feed pump of the type used for adding chlorine solution to water (Figure 29). It forms a gelatin-like mass on top of the filter bed and traps the smaller suspended particles the filter would normally let pass through the filter bed.

Maintenance consists of removing the sediment regularly by backwashing—about every 7 to 10 days. If it is allowed to build up, pump pressure will channel the water through the sediment. The water then moves at high enough velocity in the channels to carry dirt particles through the filter bed with it, and into the piping system.

A centrifuagal sand filter designed to separate particles as small as 200-mesh is available (Figure 60b). Water enters the top chamber under pressure and solids heavier than the water are forced to the perimeter of the separation chamber by centrifugal force. They drop down the side walls into the bottom while the water is drawn up through the vortex by negative pressure.

Notes

V. Determining What Pumping Installation to Make

Thus far, you have determined (1) the amount of water required for your daily needs; (2) that your water supply is safe, or can be made safe; and (3) that the water is of satisfactory quality, or the quality can be improved so that it will be satisfactory. Your next job is to determine what kind of a pumping installation will be required to meet your needs.

Your pumping installation is the heart of your water-service system. *If your pumping installation is not properly planned, you will not receive satisfactory water service.*

In this discussion, you will learn how to select the proper size and type of pump to meet your needs; how to provide ample water, even though your well or spring is a low producer; what accessories you need to protect your water system from undue wear and to provide trouble-free service; and what housing will be needed to properly protect your pump and storage tank from severe weather conditions. This information is provided under the following headings:

A. What Capacity Pump is Needed.
B. Understanding Pump Types.
C. What Type of Pump to Use.
D. What Type and Size Water Storage to Use.
E. Understanding Water-System Control Units.
F. What Housing to Provide for Pump and Water Storage.

A. What Capacity Pump is Needed?

Many home owners and some dealers are inclined to guess at what size pump is needed for a home or farm. This is very bad practice. For example, you may think that a 10-gallon-per-minute pump will supply enough water for any needs you are likely to have. But, there are few present-day installations where satisfactory water service can be supplied with a pump of that capacity.

Much of this discussion is about **"peak demand."** Peak demand is how much water your pump must deliver in gallons per minute to meet your greatest total needs whenever they occur during the day or night. For example, there may be very little, or no, demand for water in your *home* after the family retires at night. In the morning, the situation changes quickly. The pump may have to supply water to a shower, to the kitchen sink, to a clothes washer and to a toilet, all at the same time.

If you live on a *farm,* livestock watering and uses about the service buildings will add further to the peak demand. The total of all of these uses makes up the "peak demand" the pump will need to supply even though that demand may not last more than a few minutes at any one time.

If your **pump does not have enough capacity** to meet these peak demands, water pressure lowers. *Water flow decreases* to all water outlets until it gets low enough to equal the capacity of the pump to deliver water (Figure 61a). This condition may not last long. For example, if the dishwasher is causing the trouble, when it turns off, more water is supplied to the shower, to the kitchen sink and to the toilet, unless there is demand for water from some other outlet. Most people do not consider this type of water service satisfactory.

There is no way of knowing what capacity pump to purchase unless you determine the peak demand for your own particular set of conditions. With the discussion and procedures that follow, you will learn how this is done.

In determining what pumping capacity is needed for your home or farm, there are three factors to consider:

1. Capacity needed to meet peak demands.
2. Capacity needed for fire protection.
3. Rate of water yield.

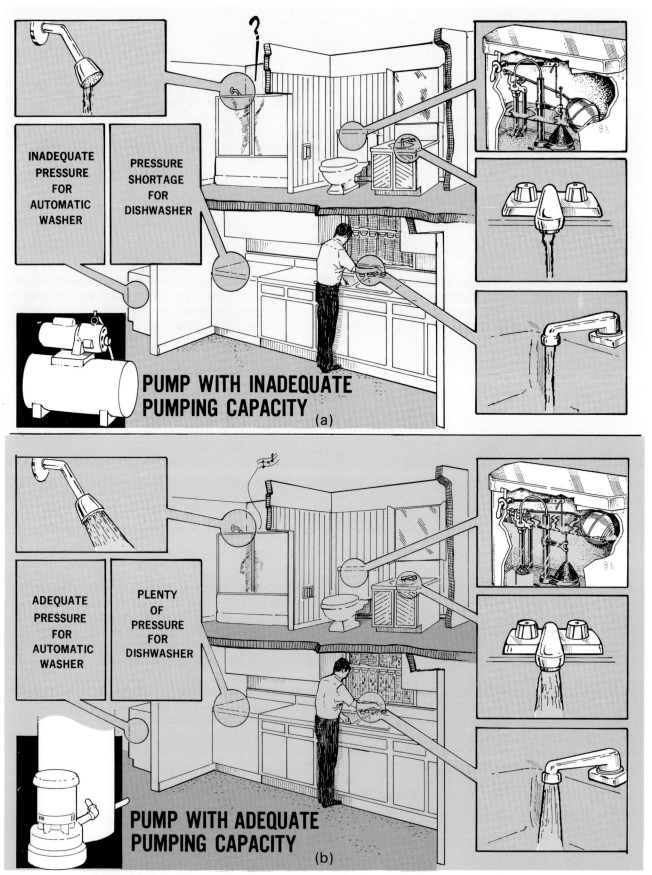

FIGURE 61. (a) A pump with too little pumping capacity to meet the water demand. (b) A pump selected to meet peak demands supplies adequate water to all water-use outlets.

CAPACITY NEEDED TO MEET PEAK DEMANDS

For many years there was no satisfactory method of determining the pump capacity needed to supply adequate water for a particular set of conditions. At best, there was considerable guessing. As the result of studies conducted by the USDA regarding water use for the home, for appliances, for watering livestock and poultry, for irrigation and for cleaning purposes outside the home, it is now possible to determine the pump capacity needed with considerable accuracy. The USDA studies show there are two kinds of water usage:

Intermittent Uses—Those which normally last for **5 minutes or less.** All general-use home applications are of this type. Seldom do any of these uses extend for longer than 5 minutes or, at most, 10 minutes. They include all household applications as well as bathroom and laundry uses.

Sustained Uses—Those that normally extend for **more than 10 minutes.** These include lawn or garden watering, livestock watering with either automatic or non-automatic waterers, and hose use for cleaning floors, automobiles or tractors.

Sustained uses are of two types: *competing sustained uses* and *non-competing sustained uses.* An example of a competing sustained use is the operation of a lawn sprinkler during the time you are using a hose for cleaning floors.

If you also plan to do automobile washing, it would be non-competing if you are the *only person* available for hose cleaning jobs. If you are cleaning floors, you would not be available to wash an automobile at the same time.

If there are *two persons* available for these jobs, then it is possible that lawn sprinkling, floor cleaning and automobile washing would all be underway at the same time. In that case, all three would be considered as competing sustained demands.

Table IX lists all of the **more common uses of water** in and about the home and farm, along with the *amount of demand* each use puts on your pump. To make it easier for you to figure pump demand, both intermittent uses and sustained uses have been interpreted in Column 1 of the table into a *"peak demand allowance"* on the pump. This figure is lower for the various uses than the actual *"individual fixture flow rate"* that could develop for each use (column 2, Table IX). The reason for the lower peak-demand figures is that studies have shown many uses do not occur at the same time. The actual demand on the pump is not the total maximum flow rate of each fixture. In fact, USDA studies of home uses show that the demand on the pump is only about ¼ of the individual fixture flow rates.

The USDA also found that most sustained demands and demands from automatic farm watering equipment develop a total demand on the pump of about ½ of the individual fixture flow rates. But, sustained uses of four hours or longer develop the same demand on the pump as the fixture-flow rate.

In arriving at a peak demand for your pump, the procedures have been broken down into three related groups of uses, each with its own set of steps. They are:
 (a) Capacity needed for household uses.
 (b) Capacity needed for irrigation, cleaning and miscellaneous.
 (c) Capacity needed for watering livestock and poultry.

If you are planning to supply water for only one or two of these groups of uses, select the group(s) that fits your needs and follow the procedures as given for each. If you plan to supply water for all three of these use groups, start with the procedures in (a) and continue through (c).

a. To determine the pump **capacity needed for household uses,** proceed as follows:

1. *List all of your home uses and the "peak demand allowance" for each.*

   Refer to Table IX, column 1.

   For example, suppose you have the following uses:

	Demand Allowance Gal. per Min.
Tub and shower	2.0*
2 lavatories (.5 gpm each)	1.0
Toilet	.75
Kitchen sink	1.0
Dishwasher	.5
Clothes washer	2.0
Water source heat pump	6.0

   *You may reduce this flow rate by installing a water-conserving shower head which reduces the flow rate.

2. *Recheck your listing for any duplicate uses or other uses you may have overlooked.*

   The example took the duplicate use into account by listing two lavatories. You may have similar duplicate uses you need to add that will affect the demand on the pump.

3. *Total the demand allowances for all household uses.*

   Based on the example in step 1, the total for that situation would be 7.25 gal. per min.

(If your pump will be supplying water to *more than one house,* use the same procedures for each additional house that you did for the first one. But, after you have figured the demand allowance for each of them, *divide the total by 2.* With each additional house, the variation in time when water is used spreads the demand on the pump over a long period of time. This keeps from adding so much to the peak demand.)

If *household uses* are all you have to figure, proceed to steps 3, 4 and 5 under (c) to determine what capacity pump is needed.

If *other water uses are involved,* proceed to (b) and/or (c). The household demand allowance you just figured will be included with the other uses.

b. To determine the pump **capacity needed for irrigation, cleaning and miscellaneous uses**—other than for livestock and poultry watering—proceed as follows:

1. *List all of the lawn, garden and miscellaneous uses you have and the water demand for each.*

As an example, suppose you have a **residence and small acreage.** Assume that your interest in additional water usage is for lawn and garden watering and a swimming pool. Use Table IX, Column 1, as your reference.

List of uses:	Demand Allowance gpm
Lawn irrigation (1 sprinkler)	2.5
Garden irrigation (1 sprinkler)	2.5
Automobile washing	2.5
Swimming pool, initial filling	2.5

If you also have **farm uses** in this group, add them to your list. Assuming the same farm situation as given on page 12 as an example, the listing might be as follows:

Cleaning milk equipment and milk storage tank	4.0
Hose-cleaning milking room	5.0
Hose-cleaning hog house	5.0
Hose-cleaning poultry house	5.0

2. *Determine from your listing which uses are competing.*

This step is important because you must use your judgment as to which demands overlap each other. You have two conditions to consider: (1) the number of self-operating units, such as sprinklers, that you will have operating at the same time, and (2) how many persons are available to operate hose-cleaning units at the same time.

In looking over the example list in step 1, suppose you decide you will *use only one sprinkler* at a time. You will either water the garden or the lawn at one time, not both at the same time. Also, you decide that no more than *two members of the family* will be using *hose-cleaning units* at any one time, one use of which is automobile washing.

This would leave the following as **competing uses:**

Lawn or garden irrigation
(1 sprinkler, self operated)
Automobile washing
(one person involved)
Heaviest hose-cleaning use
(second person involved)

The rest of the uses can then be considered as non-competing. Do not include them as part of the demand on the pump.

3. *Determine the demand allowance for the pump.*

Use Table IX as your reference for the demand figures. Using the example from step 2, the demand for this situation is figured as follows:

	Demand Allowance gpm
Lawn or garden irrigation (1 sprinkler)	2.5
Automobile washing	2.5
One hose-cleaning use (floor cleaning)	5.0
Total	10.0

If **irrigation, cleaning and miscellaneous uses** are all your pump will serve, or if they are to be *combined with the household uses* you have already figured, proceed to steps 3, 4 and 5 under (c) to determine what capacity pump is needed.

If **farm water uses** are involved, proceed to step 1 under (c).

c. To determine the pump **capacity needed for watering livestock and poultry,** proceed as follows:

1. *List all of the watering units you have in use, or plan to use in the foreseeable future, and the demand for each.*

Use Table IX, column 1, as your reference. Note that most of the table is for automatic waterers. Only the last one "watering tank" is non-automatic.

Where the term *"waterer space"* is used, *this refers to an area where only one animal can drink at a time.* Some automatic waterers have several spaces so that several animals can drink at one time.

With the open *"watering tank,"* no particular size is considered, and no maximum number of animals. Water demand is based on the use of one hand-operated valve, or an automatic float valve that may be completely open at times during heavy drinking periods.

Using the livestock and poultry numbers as indicated on page 12 as an example, the listing might look about like this:

	Demand Allowance gpm
30 dairy cattle, open housing, four watering spaces (.75 gpm each)	3.00
40 hogs, open housing, two watering spaces (.15 gpm each)	.50
1,000 laying hens, five waterers (.12 gpm each)	.60

2. *Determine the demand allowance for the pump for livestock and poultry watering.*

The total for the example in step 1 would be 4.1 gpm.

3. *Determine which use requires the greatest fixture flow.*

Check column 2, Table IX, for the greatest fixture flow for any of the water uses you have listed.

For (a) **household uses only,** the example had two uses with fixture flows of *8 gpm*—the *bath and shower* and the *clothes washer.* Use only one.

For **irrigation, cleaning and miscellaneous,** the

TABLE IX. PUMP SIZING BASED ON WATER DEMAND FOR VARIOUS USES

Water Uses	Peak Demand Allowance (for Pump)	Individual Fixture Flow Rate
	Gal. per Min.	
	(Column 1)	(Column 2)
A. Household Uses		
Bathtub, or tub-and-shower combination	2.0	(8.0)
Shower only	1.0	(4.0)
Lavatory	.5	(2.0)
Toilet—flush tank	.75	(3.0)
Sink, kitchen—including garbage disposal	1.0	(4.0)
Dishwasher	.5	(2.0)
Laundry sink	1.5	(6.0)
Clothes washer	2.0	(8.0)
Water source heat pump	8.0	(6.0)
B. Irrigation, Cleaning and Miscellaneous		
Lawn irrigation (per sprinkler)	2.5	(5.0)**
Garden irrigation (per sprinkler)	2.5	(5.0)
Automobile washing	2.5	(5.0)
Tractor and equipment washing	2.5	(5.0)
Flushing driveways and walkways	5.0	(10.0)
Cleaning milking equipment and milk storage tank	4.0	(8.0)
Hose cleaning barn floors, ramps, etc.	5.0	(10.0)
Swimming pool (initial filling)	2.5	(5.0)

Miscellaneous Uses

For items not listed, such as iron-removal filters, cow-washing systems, equipment for processing and packaging milk, liquid-feeding equipment, fogging units, egg washers, evaporative coolers, liquid manure handling units, and other special uses, determine the water demand from the manufacturer's equipment specifications. Use 50% of that figure in your listing as "Demand on pump."

C. Drinking Water for Farm Animals and Poultry*

Dairy cattle:		
Confined housing—one stanchion drinking cup for each 2 cows	.75 per cup	(1.5 per cup)
Open housing—up to 8 head per waterer space	.75 per space	(1.5 per space)
Beef cattle:		
Open housing—up to 40 head per waterer space	.75 per space	(1.5 per space)
Hogs:		
Confined housing—up to 10 pigs per waterer space	.25 per space	(0.5 per space)
Open housing—up to 25 pigs per waterer space	.25 per space	(0.5 per space)
Horses:		
Confined housing—1 watering space per 2 stalls	.75 per space	(1.5 per space)
Open lot—up to 10 horses per watering space	.75 per space	(1.5 per space)
Sheep or goats—up to 40 animals per watering space	.25 per space	(0.5 per space)
Laying hens—up to 200 birds per 4-ft. trough waterer (90°F.)	.12 per waterer	(0.25 per waterer)
Broilers—up to 100 birds per 4-ft. trough waterer (100°F.)	.12 per waterer	(0.25 per waterer)
Turkeys—up to 50 birds per fountain or per 4-ft. trough	.2 per fountain	(0.4 per fount)
Watering tank, open—for cattle, sheep & horses	2.5 per tank	(5.0 per tank)

*From USDA and Midwest Plan Service recommendations.

**Some irrigation sprinklers have more water capacity than shown in this table. If the capacity of your sprinkler is known, and it is greater than 5 gallons per minute, substitute that figure in place of 5 in the second column and ½ of that amount in the 1st column.

greatest fixture flow was for *hose-cleaning jobs—10 gpm.*

For **livestock and poultry watering,** the greatest fixture flow was 1.5 gpm for dairy cattle.

4. *Replace the demand-allowance figure for that particular use with the fixture-flow figure from column 2, Table IX.*

If you are considering the **home only,** in the example the 8 gpm would replace the 2 gpm used in the (a) listing step 1 making the total demand on the pump 13.25 gpm.

If all **three areas** are considered, the 10 gpm for floor cleaning would replace the 5 gpm used in (b), step 3. It is used rather than the 8 gpm for the home because it is the larger figure.

The reason for using this larger fixture-flow figure is to insure the pump will supply *ample water for the greatest single fixture demand.* The lesser intermittent demands for other uses are now assured of not being "robbed" when the major user has need for water.

5. *Determine what capacity pump is needed.*

Total the demand allowances for whatever uses you have listed under each group. Be sure the fixture-flow figure (step 4) has been inserted for the use requiring the greatest fixture flow.

Examples:

Home only
Tub and shower	2̶.̶0̶	8.0
2 lavatories	1.0	
Toilet	.75	
Kitchen sink	1.0	
Dishwasher	.5	
Clothes washer	2.0	
	7̶.̶2̶5̶	13.25

Home, Irrigation and Cleaning
Tub and shower	2.0	
2 lavatories	1.0	
Toilet	.75	
Kitchen sink	1.0	
Dishwasher	.5	
Clothes washer	2.0	
Lawn or garden irr.	2.5	
Automobile washing	2.5	
1 hose-cleaning use	5̶.̶0̶	10.0
	1̶7̶.̶2̶5̶	22.25

If *all three areas* are included—home, irrigation and cleaning, and livestock and poultry watering—hose cleaning still requires a larger fixture flow than any of the other uses. The 4.1 gpm demand for watering livestock and poultry is added to the 22.25 demand for the home and irrigation and cleaning uses to make a total demand on the pump of 26.35 gpm.

If you have followed these procedures accurately, you have determined the pump capacity you need to supply satisfactory water service.

CAPACITY NEEDED FOR FIRE PROTECTION

If you want your **water pump** to provide some **fire protection,** you will need at least a **10-gallon-per-minute pump** (Figure 62). Even if your peak demand figured less than 10 gallons per minute, it is well to install a pump with at least that much pumping capacity. A small fire loss could easily pay the first cost for the additional pump capacity.

According to the National Fire Protection Association:[31]

Any appraisal of the need for water for fire protection requires a knowledge of "fire load."

Fire load refers to the amount of combustible material present in a given situation, expressed in pounds per square foot. Rural fire loads are exceedingly variable and thus the size of hose stream required is equally variable. For instance, from 2½ to 10 gallons per minute may be an adequate first-aid stream for dwellings, whereas a sizable wooden barn filled with hay that had a pre-burn of 25 minutes may require 1,000 gallons per minute.

FIGURE 62. For fire protection, a pump with a capacity of at least 10 gallons per minute is needed. It will not extinguish a fire that is well under way, but it will help protect adjacent buildings from catching on fire.

If your pump has a capacity of **10 gallons per minute,** it will supply enough water for a ¼-**inch fire nozzle.** At 30 pounds (psi) pressure the nozzle will produce a stream that will reach 30 feet high, or a stream that will extend horizontally about 40 feet. This is not enough water to be effective with a large fire, or a fire that is well under way, but it can be very effective in either putting out or controlling a small fire. Sometimes, it can be most effective if used in keeping another building dampened down so as to keep it from catching fire (Figure 62).

If you install a pump with as much as **25 gallons per minute,** there will be enough water for **one ⅜-inch nozzle or two ¼-inch nozzles.** With a ⅜-inch nozzle at 30 pounds (psi) pressure, you can provide a stream that will reach 40 feet high or extend horizontally about 65 feet.

If you wish more protection than that which your water pump will provide, you may need to install a reserve storage tank where 3,000 to 5,000 gallons of water can be stored continuously for use by community fire-fighting equipment. This will be discussed later under "D. What Type and Size Water Storage to Use."

RATE OF WATER YIELD

The next factor you will need to consider is whether your water source will supply water as fast as your pump will need it. This applies mostly to springs or wells. You determined the rate of yield of your water source when you measured the "Amount of Water Available," page 16.

If it is evident your water source will supply enough water to take care of your peak demand, you can use a pump of the same capacity that you determined meets your peak demand. The water can be pumped directly from the source.

If your peak demand is more than the hourly yield of your water source, you will still need the same size of pump to provide water to meet your periods of high demand, but the extra water will have to be supplied from a water storage. A **second pump** will be needed to provide water for the storage from the source. This pump will need to be sized to match the capacity of the source to supply water. This is also discussed under "D. What Type and Size Water Storage to Use."

B. Understanding Pump Types

Now that you have determined the amount of water you need to meet your peak demands, you are ready to select a pump that provides that much capacity, or more. But, before selecting a pump, you need to know how the various pumps work. In doing so, you will learn their names and gain an understanding of their limitations and capabilities. With this information it will be easier for you to work with a pump dealer in securing the type of pump that will best fit your needs.

The information is discussed under the following headings:
— Principles involved in pumping
— How a piston pump works
— How a centrifugal pump works
— How a centrifugal-jet (ejector) pump works
— How a turbine pump works

This discussion deals only with the **water end**—pumping mechanism—of a pump. The **power end** —motor-and-drive mechanism—is determined by the manufacturer. Each pump is limited in the amount of water it will deliver and the pressures under which it will deliver the water. Consequently, each manufacturer selects a motor and drive that will be satisfactory under the range of operating conditions for which a particular pump model is designed.

PRINCIPLES INVOLVED IN PUMPING

There is a wide assortment of pumps for home and farm use. But, no matter how they are made, they do a certain amount of **pulling water by suction** and a lot of **pushing of water by pressure** (Figure 63). The

FIGURE 63. The job of a pump is to lift water by suction into its mechanism, then deliver the water under pressure to a storage or pressure tank.

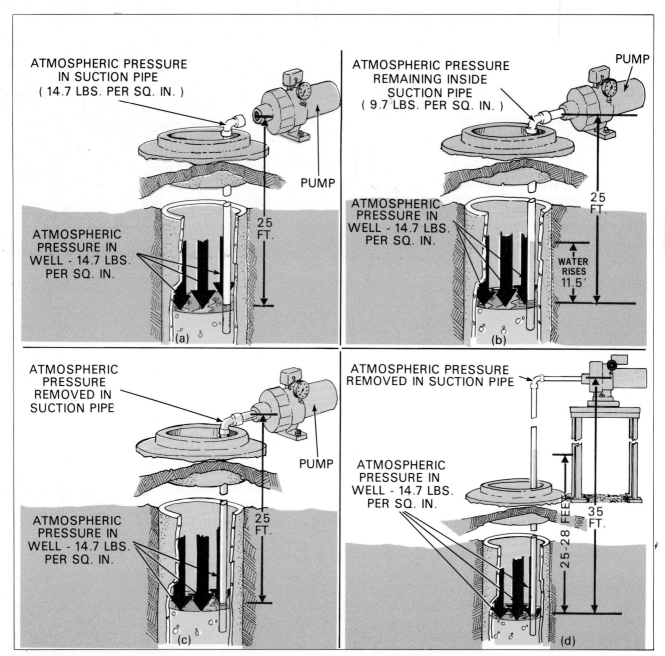

FIGURE 64. How a pump lifts water by suction. (a) With the pump suction pipe disconnected, atmospheric pressure on the water in the well and the water inside of the pipe is the same—14.7 pounds per square inch (at or near sea level). (b) When the suction pipe is connected to the pump and the pump is started, air pressure in the suction pipe is reduced (a partial vacuum develops) and the water rises from the well into the pipe. (c) As pumping action continues, all air is pumped from the suction pipe so that atmospheric pressure pushes the water into the pump. (d) If the pump is raised too high above the water source, there is not enough atmospheric pressure to push the water into the pump.

suction comes when your pump draws water into its working mechanism from your water source. Then the pump puts the water under pressure as it forces the water into the pressure tank. From the tank, it flows under pressure through your piping system.

The **height water can be lifted by suction**—from the water source to the level of the pump—**is quite limited.** Even the best suction pumps cannot lift water more than about 22 to 25 feet straight up on a clear day at sea level. Some are limited to 15 feet or less. Here is the reason why. When water is lifted by the suction, all the pump does is remove part or all of the atmospheric pressure—develop a partial or complete vacuum—in the suction pipe.

Note in Figure 64a, the pump is not connected to the suction pipe. Atmospheric pressure inside the pipe and in the well is the same—14.7 pounds per square inch (psi).

When the pump starts (Figure 64b), it **removes more and more air** from the suction pipe. This gradually reduces the air pressure inside the pipe while the atmospheric pressure on the outside remains the same—14.7 psi. Water from the well is pushed into the suction pipe because of this difference in pressures. As the pump continues to operate, more and more air is removed—creating more and more of a vacuum effect—until the atmospheric pressure finally forces the water up to the level of the pump (Figure 64c). *With all air removed from the suction pipe,* the pump can continue to operate and take advantage of the action of the atmospheric pressure lifting the water to its level.

If a **leak develops in the suction pipe,** air enters. The vacuum effect is destroyed, and the water level drops in the suction pipe down to the same level as that of the water source. When this happens the pump is said to "have lost its prime."

Assume the pump is moved to a height of 35 feet above the water level in the well (Figure 64d), and the pumping action started. With most pumps, the water would be lifted about 15 to 28 feet high depending on the type of pump. (If you were able to take full advantage of atmospheric pressure, you could lift the water as high as 33.9 feet, but pumps are not that efficient.) The space in the suction pipe between the 25- to 28-foot level and the 35-foot level would remain empty, except for a small amount of air and water vapor—no matter how long you pump. There is not enough vacuum for the atmospheric pressure to lift the water any higher.

The figure of **14.7 pounds (psi) is atmospheric pressure at sea level.** Atmospheric pressure becomes less and less as you rise above sea level. If you live at a high elevation, subtract 1 foot from the pump suction lift for each 1,000 feet of elevation. For example, if you live 2,000 feet above sea level, a pump that lifts water 25 feet by suction at sea level will lift it only 23 feet at the higher elevation when in proper condition.

You cannot depend on a pump lifting water by suction more than 25 feet. A piston pump, in good mechanical condition, located not more than a few hundred feet above sea level, may be able to do so. There are other types that can not. You occasionally hear of someone who claims he has a pump that is lifting water by suction as much as 30 or 35 feet. The 30-foot lift is highly improbable. The 35-foot lift is impossible—atmospheric pressure will not lift water that high.

Pumps that lift water by suction from these limited depths are called "*shallow-well*" *pumps.*

Pumps that lift water from greater depths are called "*deep-well*" *pumps.* Pumps of this type provide for lowering part, or all, of the pumping mechanism down into the well to a point where little or no suction is needed. The "no-suction" condition is when the pumping mechanism is submerged in the water (Figure 68).

HOW A PISTON PUMP WORKS

A **piston pump** was one of the earliest automatic pumps developed for home and farm use. Its operating principle is basically the same as the once popular hand-operated piston pump. Here is how the hand-operated pump works.

As you push down on the handle of a hand-operated piston pump (Figure 65a), water is lifted by the plunger, and it flows out of the spout. At the same time, a partial vacuum develops below the plunger. The partial vacuum causes the water in the suction pipe to force open the check valve and fill the cylinder below the plunger.

As you lift the pump handle (Figure 65b), the water below the plunger is trapped by the closed check valve in the bottom of the cylinder. At the same time, a valve in the plunger lifts and lets the water pass to the upper side of the plunger. It is then ready for discharge when the pump handle is pushed down and the plunger is raised again. This is called a *"single-acting" piston pump.* This is the simplest type of shallow-well suction-type pump.

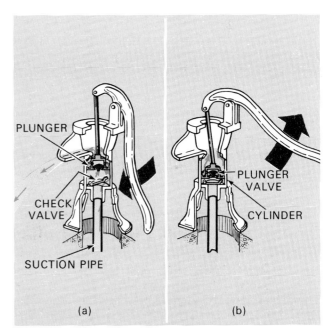

FIGURE 65. How a hand-operated piston pump works. (a) As the handle is pushed down, causing the plunger to rise, water is discharged through the pump spout. The vacuum that develops in the cylinder below the plunger is filled with water from the suction pipe which connects the pump to the water source. (b) As the pump handle is raised, the plunger lowers. This causes the check valve to seat and trap water in the cylinder. The trapped water forces the plunger valve open allowing the water to enter the cylinder area on the upper side of the plunger.

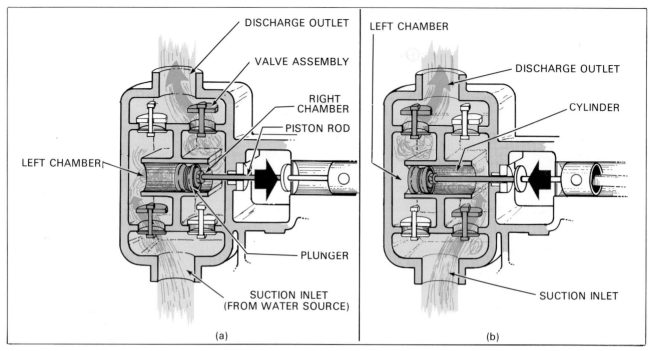

FIGURE 66. How a double-acting piston pump works. (a) Movement of the plunger to the right pulls water from well into the left chamber and forces water out of right chamber. (b) Plunger movement to left forces water out of left chamber and pulls water from the well into the right chamber.

If you purchase a motor-driven **shallow-well piston pump,** it will be one that works on the principle you just studied, but is improved by use of the **double-acting principle.** That is, it pulls water from the well on one side of the plunger at the same time it forces water out on the other side (Figure 66a). On the return stroke (Figure 66b), the action is reversed—the side that was pulling water into the cylinder now forces it out. The side that was forcing water out is now pulling water into the cylinder. This is the reason for calling it a "double-acting" pump. The double-acting design makes a close-coupled compact unit, one that will work at a fairly high speed.

Figure 67 shows what a double-acting shallow-well pump looks like.

If the water in your well is more than 22 to 25 feet below pump level, a **deep-well piston pump** is used. The plunger and cylinder are put down into the well to within a few feet of water level or, better yet, below water level (Figure 68). It works on exactly the same principle as the hand-operated pump (Figure 65). This type of pump works at a fairly slow speed (about 45-65 strokes per minute) because of the long sucker rod that extends from the pump head down to the cylinder to work the plunger.

All piston pumps are called *"positive acting"* because each stroke of the plunger moves a constant amount of water. Since it is positive acting, it **will pump air** from the suction line without any difficulty. A small air leak in the suction line simply slows down the amount of water delivered.

FIGURE 67. A double-acting shallow-well piston pump.

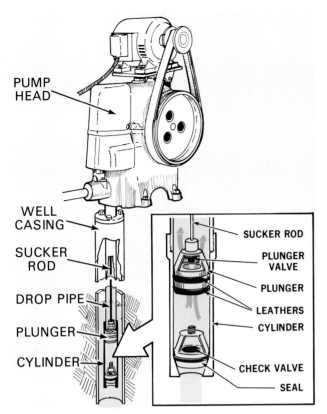

FIGURE 68. A deep-well piston pump is installed with the pumping mechanism in the well, usually below water level.

HOW A CENTRIFUGAL PUMP WORKS

A **centrifugal pump** is of very simple design. The only moving part is an impeller wheel attached directly to the motor shaft.

To understand how an **impeller** develops pumping action, let us examine a very simple centrifugal pump. In Figure 69a, the impeller consists of a suction pipe and lateral arm. It is turned by an electric motor. As long as there is air in the pipe, and in the lateral arm, there can be no pumping action no matter how fast it is turned.

If the suction pipe and lateral arm are filled with water and then started rotating, water in the lateral arm is thrown out by centrifugal force. This creates a partial vacuum which lifts more water from the bucket. In this way, the pumping action continues as long as the assembly spins and no air enters it.

If you add more lateral arms to the pipe, more water is pumped (Figure 69b). These arms provide the same action as the vanes in a centrifugal pump impeller (Figure 69c). The space between each pair of vanes acts in the same way as the lateral arms.

The pumps shown in Figure 69a and b are not useable, partly because the suction pipe turns with the rotating lateral arms. **Factory-built centrifugal pumps** are equipped with a **wearing ring** on the suction inlet next

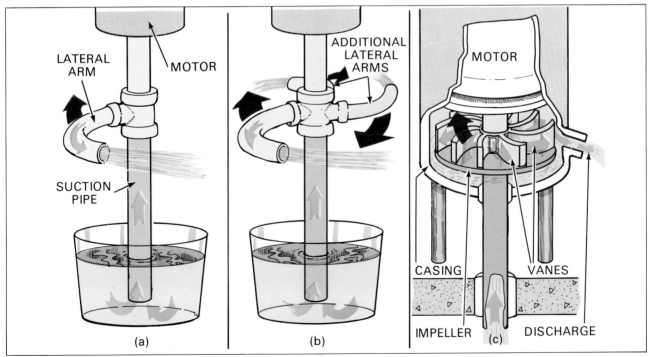

FIGURE 69. How a centrifugal pump works. (a) An L-shaped pipe, completely filled with water and rotated rapidly, will pump water out of a bucket. Water thrown out of lateral arm by centrifugal force creates a suction, causing water to rise from the bucket. (b) By adding more lateral arms, more water is pumped. (c) With a manufactured pump, the lateral arms are replaced with an impeller mounted inside of a casing. The impeller vanes are mounted with a plate on one side (as shown) or with a plate on each side. These vanes act in the same way as the lateral arms in (b).

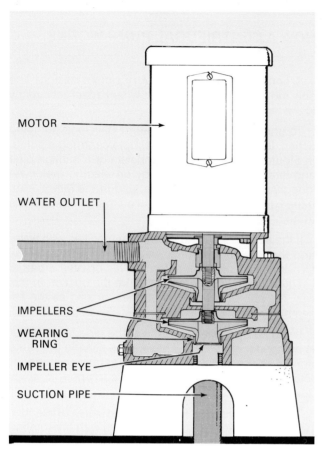

FIGURE 70. When water must be supplied at higher pressures, two or more impellers are used on a centrifugal pump.

to the impeller. The **eye** of the impeller is machined to fit very closely into the wearing ring (Figure 70). The only moving part is the impeller.

A centrifugal pump is not positive acting. As the *water level lowers* in your well, it pumps less and less water. Also, when it pumps against *increasing pressure*, it pumps less and less water. For these reasons, it is important in selecting a centrifugal pump that you get one designed to do your particular pumping job.

For higher pressures or greater lifts, two or more impellers are commonly used to meet the needs, as shown in Figure 70.

Shallow-well centrifugal pumps are seldom used if the **suction lift is more than about 15 to 18 feet.** This has limited their use for shallow-well pumping, unless they are equipped with an ejector. The combination centrifugal pump and ejector is another type of pump which will be studied under the next heading.

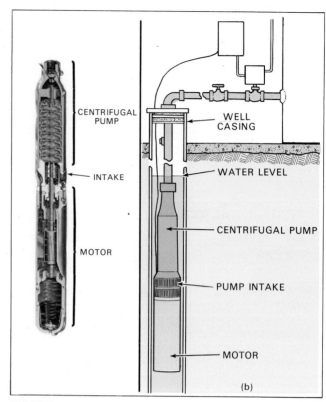

FIGURE 71. (a) Sectional view of a deep-well submersible centrifugal pump. (b) Submersible pump and motor is installed in a well below water level.

Deep-well centrifugal pumps have become quite popular. They are known as "submersible" or "submergible" pumps. Note in Figure 71a that a submersible unit consists of several impellers driven by a motor that attaches below it. Pumps with more than one impeller are sometimes called "multi-stage" pumps. The number of impellers is determined by how far the water has to be lifted and how much pressure is needed at the point of delivery. Each time water is pumped from one impeller to the next one, its pressure is increased. This is a popular pump because of positive action and having no need for freeze protection.

The motor-and-pump assembly are closely connected and built as a unit. The motor is specially designed for use under water. The whole assembly, including the motor, is let down into the well on the end of a drop pipe to a position below the water level. In this way, the pumping mechanism is always filled with water (primed) and ready to pump (Figure 71b).

Centrifugal pumps, both the shallow-well type and the deep-well type, have either **little or no ability to pump air.** When starting, the pump and suction line needs to have all of the air removed. An air leak in the suction line will cause the pump to quit pumping.

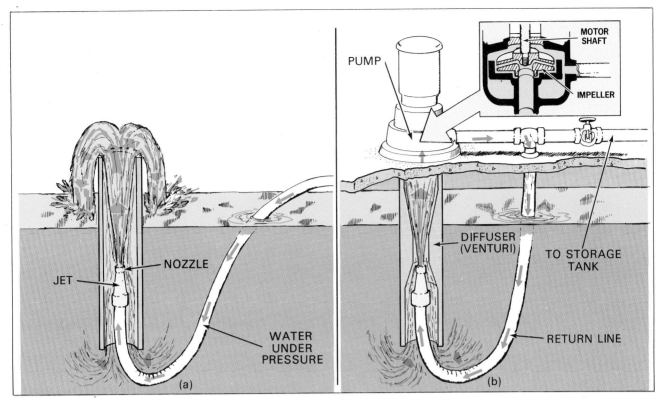

FIGURE 72. (a) How a jet provides pumping action. Water is supplied to the jet nozzle under pressure. Water surrounding the jet stream is lifted and carried up the pipe as a result of the jet action. (b) When a jet is used with a centrifugal pump, a portion of the water delivered by the pump is returned to the jet nozzle to operate it. The jet lifts water from the well to a level where the centrifugal pump can finish lifting it by suction.

HOW A CENTRIFUGAL-JET (EJECTOR) PUMP WORKS

The **addition of a jet to a centrifugal pump** makes it adaptable to many more conditions. But first you need to understand what a jet is and how it works.

A **jet** is a nozzle which receives water at high pressure (Figure 72a). As the water passes through the jet, water speed (velocity) is greatly increased, but the pressure drops. This action is the same as the squirting action you get with a hose when you start to close the nozzle. The greatly increased water speed, or squirting action, plus the low pressure around the tip of the nozzle, is what *causes suction to develop* around the jet nozzle. Air, water or loose materials around a jet nozzle are drawn into the water stream and carried along with it.

If the squirting action is confined inside a pipe (Figure 72a), the high-speed stream will suck in additional water and carry it along, thus providing pumping action.

For a jet to be efficient and effective in a combination centrifugal-jet pump, it must be used with a cone-shaped member called a **"diffuser"** or **"venturi"** (Figure 72b). In this discussion it will be called a diffuser, since it is widely known by that name in the pump industry. The diffuser changes the high-speed jet stream back to a high-pressure stream for delivery to the centrifugal pump.

The **jet and the diffuser** are simple in appearance, but they have to be well engineered and carefully matched to be efficient for various pumping conditions. Pumps of this type are usually called "jet pumps". Also, the jet-and-diffuser combinations are known as *"ejectors."* Those terms will be used in the rest of this discussion.

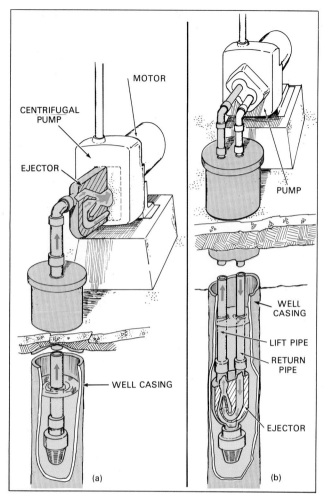

FIGURE 73. (a) With a shallow-well jet pump, the ejector is mounted close to the pump impeller. (b) With a deep-well jet pump, the ejector is usually mounted just above the water level in the well, or else submerged below water level.

Many shallow-well jet pumps are built so the ejector may be removed from the pump casing (Figure 73a) and connected to the drop pipe in the well (Figure 73b), thus making it possible to *convert a shallow-well jet pump into a deep-well jet pump.* This will probably require a change of ejectors to work efficiently. This is of particular value when you have a water level that is gradually lowering in your water source.

Jet pumps have the same **air-handling** characteristics as the centrifugal pump that makes up the unit. In general, the pump has to be started with the pump and piping connections to the water supply completely filled with water.

HOW A TURBINE PUMP WORKS

Turbine pumps look very much like centrifugal pumps, and they operate in much the same manner. However, they are not as dependent on centrifugal force as the straight-centrifugal pump. Although there is some centrifugal action, there is also propelling or paddling action to provide a lifting effect.

Figure 74 shows how a **shallow-well turbine pump** works. It uses one impeller. The design is such that it spins the water several times as it passes from the pump inlet to the outlet. At the same time, the fins provide some pushing action. This improves its suction over that of a straight centrifugal pump. It will lift water by suction from *depths of 25 feet* or more. It can also **pump some air.**

Deep-well turbine pumps are usually of the submersible type. They look very similar to the centrifugal-submersible pump shown in Figure 71.

With a **shallow-well jet pump,** the ejector is located next to the impeller (Figure 73a). A portion of the water from the impeller is forced through the ejector, and the rest is delivered to the pressure tank. With the ejector located on the suction side of the pump, the *suction is increased* considerably. This enables a centrifugal pump to increase its effective suction lift from about 15 feet to as much as 25 feet. But, the amount of water delivered to the storage tank becomes less as the distance from the pump down to the water increases—more water has to be returned to operate the ejector.

The difference between a **deep-well jet pump** (Figure 73b) and a shallow-well jet pump is the location of the ejector. The deep-well ejector is located in the well below water level. (The ejector design varies for different depths, but the manufacturer takes care of that problem.) The deep-well ejector works in the same way as the shallow-well ejector. Water is supplied to it under pressure from the centrifugal pump (Figure 73b). The ejector then returns the water, plus an additional supply from the well, to a level where the centrifugal pump can lift it the rest of the way by suction.

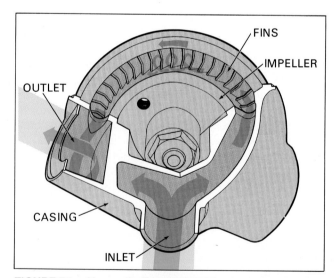

FIGURE 74. How a shallow-well turbine pump works. The fins on the impeller give the incoming water a spinning action. The water spirals from the tips of the fins, into the raceway and back into the fins, several times before it reaches the outlet. Each time it is thrown from the fin tips, its pressure is increased.

C. What Type of Pump to Use

With the information you developed on the pump capacity you will need, and with a knowledge of how the different pumps work, you are now in position to decide on what type of pump to use. The one you select will be determined by the following factors:
— Depth to water
— Well size
— Pressure range needed for adequate water service
— Height water is lifted above pump
— Pump location
— Pump durability and efficiency
— Dealer service

You will find **Tables X, XI and XII** of help in dealing with the first three factors. The tables **give ranges of pumping capacities** for the different pump types operating under the various conditions listed. You may wonder how accurate the manufacturers' data is for these pumping capacities. The national Water Systems Council (manufacturers of jet, submersible and reciprocating pumps) has a certification program to help provide this assurance. Any pump with a WSC *"Certified Performance"* sticker on it is guaranteed by the manufacturer to produce within 10 per cent of its rated capacity while new and operating under the conditions for which it is designed.

DEPTH TO WATER

If the water in your well or water source never gets **lower than 15 to 25 feet below the pump,** you can use a **shallow-well pump**—one of the types you just finished studying under "Understanding Pump Types." Compared to deep-well pumps, they cost less to buy, and connecting the pump to the water source is simple. All your plumber needs to do is run a suction pipe from your pump to the water source (Figure 73a).

You have already determined how much water you will need per hour to meet your peak demands. Now you need to know which of the **pump types** will supply that quantity of water and lift it as high as needed to reach your water-use outlets. Tables X and XI show these various pump types and the range of pumping capacities for each type at different depths. Pump manufacturers publish the capacities of their pumps when pumping from various depths, so there is no need for your dealer to guess at the capacity his pumps will deliver under your conditions.

If the water in your well or water source is **more than 25 feet below pump level,** Table XII shows the types of deep-well pumps available to you and their capacities at various depths.

If you have a **shallow well, but you have observed the water level lowering** over a period of several years, the jet pump may best fit your needs. As shown in Figure 73, some of these are designed for changing from shallow- to deep-well use when the water level lowers.

WELL SIZE

When you have found the pump type that will lift water from your water source to pump level and provide the amount of water you need, your next check is to see if it will work with **your size of well.** Note that in all three tables the different well sizes are given in the left column for the different pump types along with their capacities.

Well size is mostly a **problem for deep-well installations.** Ones less than 4 inches in diameter are the biggest problem, especially if you have need for large amounts of water. Manufacturers of piston pumps and jet pumps have tried to meet this problem with special cylinders and ejectors.

A **deep-well piston pump** can be used on a well as small as 2 inches in diameter. This can be accomplished

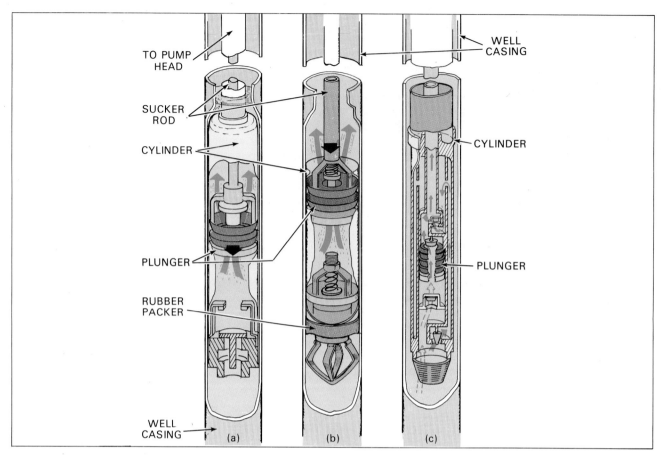

FIGURE 75. Piston-pump cylinders for extra small wells (2-4 inches in diameter). (a) Single-acting closed-type cylinder. (b) Single-acting Eureka cylinder. Packer at bottom of cylinder makes water-tight connection so well casing may be used in place of drop pipe. (c) Double-acting cylinder. Pumps water on both up and down strokes.

with a small, *closed-type single-acting cylinder* (Figure 75a), a *Eureka cylinder* (Figure 75b), or a *double-acting cylinder* (Figure 75c). The latter two will deliver more water than the former. Table XII shows the comparative pumping capacity for each of these cylinders from various depths and with various well sizes.

The **Eureka cylinder** (Figure 75b) is equipped with a water-tight rubber packer that fits against the cylinder wall. This does away with a drop pipe for connecting the cylinder to the pumping head and provides room for a larger cylinder.

A **double-acting deep-well cylinder** (Figure 75c), in the same size well as the other two, delivers more water than either of the others. It pumps water on both the up stroke and down stroke.

Any of these extra-small cylinders will probably have to be ordered as special equipment by your dealer.

Jet pumps can be used with a well size as small as 2 inches. If your well casing is in good condition, you can use it in place of one pipe. To do this requires a packer-type ejector to keep water in the upper section of the casing from passing down into the well (Figure 76). In this way, the casing acts as a return pipe. Water is forced down the casing from the pump and into the ejector. The ejector then acts in the usual manner in lifting water

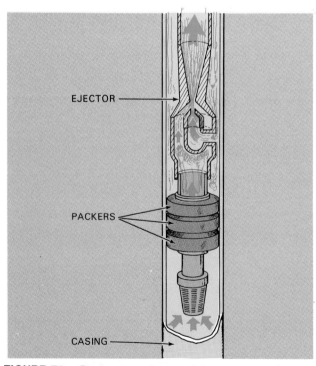

FIGURE 76. Packer-type jet used for extra-small wells (2 to 4 inches in diameter). Also used for larger wells where maximum pumping capacity is needed.

TABLE X. PUMP SELECTION CHART—SHALLOW WELLS, LOW PRESSURE*

For pumps lifting water from depth of **25 feet or less** and delivering to faucets **not more than 20 feet above level of pump**.
Capacities based on delivery against **pounds-per-square-inch tank pressure**.

Well Size	Total Lift (Distance from lowest water level to pump)	Piston Pump	Shallow-well Jet	Shallow-well Turbine	Straight Centrifugal**
		\multicolumn{4}{c}{Range in Pump Capacities (gal. per hr.)}			
1¼″ Diameter					
	10 feet	250 — 500	400 — 1830	400 — 565	450 — 684
	15 feet	250 — 500	375 — 1650	390 — 555	350 — 672
	20 feet	250 — 500	285 — 1440	380 — 545	275 — 646
	25 feet	250 — 500	240 — 780	370 — 535	
1½″ Diameter					
	10 feet	250 — 500	440 — 2800	400 — 1330	450 — 2300
	15 feet	250 — 500	360 — 2640	390 — 1310	350 — 2200
	20 feet	250 — 500	285 — 2500	380 — 1290	275 — 2050
	25 feet	250 — 500	210 — 1200	370 — 1270	200 — 1900
2″ Diameter (or larger, also includes dug wells, cisterns, springs, ponds and lakes)					
	10 feet	250 — 500	440 — 3660	400 — 1330	450 — 3500
	15 feet	250 — 500	360 — 3540	390 — 1310	350 — 3300
	20 feet	250 — 500	285 — 3420	380 — 1290	275 — 3100
	25 feet	250 — 500	210 — 3180	370 — 1270	200 — 1900

*The range of pumping capacities in this table are not for any one pump, but rather for a number of pumps, with different capacities, operating under the conditions given. All figures are taken from manufacturers' published ratings on pumps. If more pumping capacity is needed, check Table XI. Some high pressure pumps have more capacity than low-pressure pumps and can often replace them.

**Many manufacturers of straight centrifugal pumps recommend limiting their use to 15 feet or less.

from the well back up the center pipe to the pump. In some pumps, water flow is the reverse of that just described. Water is forced from the pump down the center pipe to the ejector. This type must be located directly over the well and normally requires more maintenance than other types.

TABLE XI. PUMP SELECTION CHART—SHALLOW WELL, HIGH PRESSURE*

For pumps lifting water from depths of 25 feet or less and delivering to faucets higher than 20 feet above pump level.
Capacities based on **20 pounds per square inch of tank pressure at delivery level.**

Well Size	Total Lift (Distance from lowest water level to pump)	Shallow-well Piston Pumps — Height Pump to Delivery Point (25-90 ft. / to 140 ft. / to 200 ft.)			Centrifugal Single and Multistage Straight Centrifugal — Height Pump to Delivery Point (25 ft. / 40 ft.)		Shallow-well Turbine — Height Pump to Delivery Point (25 ft. / 100 ft. / 200 ft.)			Shallow-well Jet — Height Pump to Delivery Point (25 ft. / 40 ft. / 90 ft.)		
		\<Range in Pump Capacities (gal. per hr.)\>										
1¼″ Dia.	10 feet	260- 540	260- 580	260- 580	900-1400	900-1200	0- 227			530-1450	345-1080	345- 450
	15 feet	260- 540	260- 580	260- 580	750-1200	750-1100	0- 220			460-1270	300- 930	300- 400
	20 feet	260- 540	260- 580	260- 580	600-1000	600-1000	0- 213			370-1080	240- 750	240- 350
	25 feet	260- 540	260- 580	260- 580	450- 800	450- 300	0- 206			240- 780	200- 540	200- 300
1½″ Dia.	10 feet	260-1020	260-1020	260-1020	2200-2500	1800-2100	227-1240	270-1460	285-1120	530-2610	345-1080	345- 450
	15 feet	260-1020	260-1020	260-1020	1950-2300	1450-1850	220-1220	265-1440	282-1115	460-2280	300- 930	300- 400
	20 feet	260-1020	260-1020	260-1020	1800-2100	1200-1600	213-1200	260-1425	279-1110	370-1920	240- 750	240- 350
	25 feet	260-1020	260-1020	260-1020	1600-2000	1000-1500	206-1180	255-1410	276-1105	240-1260	200- 540	200- 540
2″ Dia.	10 feet	260-1680	260-1680	260-1020	2200-5600	1800-2100	227-1240	270-1460	285-1120	530-2610	345-1080	810-3350
	15 feet	260-1680	260-1680	260-1020	1950-5250	1450-1850	220-1220	265-1440	282-1115	460-2280	300- 930	600-2800
	20 feet	260-1680	260-1680	260-1020	1800-2100	1200-1600	213-1200	260-1425	279-1110	370-1920	240- 750	420-2340
	25 feet	260-1680	260-1680	260-1020	1600-2000	1000-1500	206-1180	255-1410	276-1105	240-1260	200- 540	0-1700
2½″ Dia.	10 feet	260-2640	260-2640	260-1020	2200-5600	1800-2100	227-1240	270-1460	285-1120	530-2610	345-1080	810-3350
	15 feet	260-2640	260-2640	260-1020	1950-5250	1450-1850	220-1220	265-1440	282-1115	460-2280	300- 930	600-2800
	20 feet	260-2640	260-2640	260-1020	1800-2100	1200-1600	213-1200	260-1425	279-1110	370-1920	240- 750	420-2340
	25 feet	260-2640	260-2640	260-1020	1600-2000	1000-1500	206-1180	255-1410	276-1105	240-1260	200- 540	0-1700
3″ Dia. (or larger, also includes dug wells, cisterns, ponds and lakes)	10 feet	260-3960	260-2640	260-2640	2200-5600	1800-2100	227-1240	270-1460	285-1120	530-2610	345-1080	810-3350
	15 feet	260-3960	260-2640	260-2640	1950-5250	1450-1850	220-1220	265-1440	252-1115	460-2280	300- 930	600-2800
	20 feet	260-3960	260-2640	260-2640	1800-2100	1200-1600	213-1200	260-1425	249-1110	370-1920	240- 750	420-2340
	25 feet	260-3960	260-2640	260-2640	1600-2000	1000-1500	206-1180	255-1410	246-1105	240-1260	200- 540	0-1700

*The range of pumping capacities in this table are not those of any one pump, but rather for a wide range of pumps with different capacities made by different manufacturers but operating under the conditions given. All figures are taken from manufacturers' published ratings on pumps offered for individual water supplies.

TABLE XII. PUMPS FOR LIFTING WATER FROM MORE THAN 25 FEET DEPTH

Water sources include driven, drilled, bored and dug wells.
Capacities based on **20 lbs. per square inch tank pressure at point of delivery.**

| Well Size | Total lift (Lowest water level to highest delivery point except for jets) | Piston (plunger) Pumps | | | Centrifugal Jet | | | | | | Centrifugal Submersible (Gals. per hour) | Deep-well Turbine (Gals. per hour) |
| | | | | | Pump and building on same level | | 50 ft. | | 75 ft. | | | |
		Single-Acting Cylinder	Double-Acting Cylinder	Eureka Cylinder*	1-Pipe	2-Pipe	1-Pipe	2-Pipe	1-Pipe	2-Pipe		
2" Diameter	30 feet	170- 225	300- 385	190- 270	310- 900	Not adaptable	270-1032	Not adaptable	250-1000	Not adaptable	Not adaptable	Not adaptable
	50 feet	170- 225	300- 385	190- 270	245- 620		220- 690		190- 790			
	70 feet	170- 225	300- 385	190- 270	180- 500		160- 546		140- 650			
	100 feet	170- 225	300- 385	190- 270	165- 350		0- 315		120- 490			
	125 feet	170- 225	300- 385	190- 270	140- 250		0- 220		110- 405			
	150 feet	170- 225	300- 385	190- 270	120- 180		0- 192		0- 360			
	200 feet	170- 225	300- 385	190- 270	100- 125		0- 100		0- 100			
	250 feet	170- 225	300- 385	190- 270								
	300 feet	170- 225	300- 385	190- 270								
	350 feet	170- 225	300- 385	190- 270								
2½" Diameter (2½" smallest well using open type cylinder)	30 feet	180- 250	300- 445	310- 465	400-1480	Not adaptable	360-1380	Not adaptable	320-1020	Not adaptable	Not adaptable	Not adaptable
	50 feet	180- 250	300- 445	310- 465	360-1270		310-1070		290- 850			
	70 feet	180- 250	300- 445	310- 465	210- 970		200- 870		170- 700			
	100 feet	180- 250	300- 445	310- 465	200- 600		180- 540		160- 430			
	125 feet	180- 250	300- 445	310- 465	180- 450		100- 400		140- 320			
	150 feet	180- 250	300- 445	310- 465	170- 400		150- 260		130- 260			
	200 feet	180- 250	300- 445	310- 465	160- 240		140- 220		120- 170			
	250 feet	180- 250	300- 445	310- 465	150- 200		130- 180		100- 140			
	300 feet	180- 250	300- 445	310- 465	140- 150		120- 130		95- 100			
	350 feet	180- 250	300- 445	310- 465								
3" Diameter	30 feet	180- 285	300- 480	465- 625	400-2250	350- 470	360-2030	0- 582	280-1800	0- 576	Not adaptable	Not adaptable
	50 feet	180- 285	300- 480	465- 625	360-1900	280- 330	330-1700	0- 384	260-1640	0- 444		
	70 feet	180- 285	300- 480	465- 625	250-1600		220-1400	0- 240	170-1120	0- 384		
	100 feet	180- 285	300- 480	465- 625	180-1200		160-1100		130- 880	0- 252		
	125 feet	180- 285	300- 480	465- 625	160- 900		140- 810		110- 650	0- 175		
	150 feet	180- 285	300- 480	465- 625	150- 750		130- 670		100- 520			
	200 feet	180- 285	300- 480	465- 625	140- 500		0- 450		0- 360			
	250 feet	180- 285	300- 480	465- 625	130- 330		0- 290		0- 230			
	300 feet	180- 285	300- 480	465- 625	120- 230		0- 210		0- 170			
	350 feet	180- 285	300- 480									
3½" Diameter	30 feet	180- 360	300- 720	Requires special cylinder		Not adaptable		Not adaptable		Not adaptable	Not adaptable	Not adaptable
	50 feet	180- 360	300- 720									
	70 feet	180- 360	300- 720									
	100 feet	180- 360	300- 720									
	125 feet	180- 360	300- 720		Same as for 3" well		Same as for 3" well		Same as for 3" well	Same as for 3" well		
	150 feet	180- 360	300- 720									
	200 feet	180- 360	300- 720									
	250 feet	180- 360	300- 720									
	300 feet	180- 360	300- 720									
	350 feet	180- 360	300- 720									

*Minimum capacity based on 6" stroke at 50 strokes per minute. Maximum based on 9" stroke at 45 strokes per minute.
**Larger units can be built to fit individual needs.

TABLE XII. (CONTINUED) PUMPS FOR LIFTING WATER FROM MORE THAN 25 FEET DEPTH

Water sources include driven, drilled, bored and dug wells.
Capacities based on **20 lbs. per square inch tank pressure at point of delivery.**

Well Size	Total lift (Lowest water level to highest delivery point except for jets)	Piston (plunger) Pumps			Centrifugal Jet						Centrifugal Submersible (Gals. per hour)	Deep-well Turbine (Gals. per hour)
					Height — Pump to Delivery Point							
					Pump and building on same level		50 ft.		75 ft.			
		Single-Acting Cylinder	Double-Acting Cylinder	Eureka Cylinder*	1-Pipe	2-Pipe	1-Pipe	2-Pipe	1-Pipe	2-Pipe		
		Range in Pump Capacities (gal. per hr.)										
4" Diameter	30 feet	180- 585	300- 720	675-1100	480-3000	400-1400	450-1700	360-1260	450-4000	320-1120	640-4000	2160-7860
	50 feet	150- 585	300- 720	675-1100	450-1900	330-1000	400-1700	290- 900	410-3350	260- 800	570-3600	1560-7560
	70 feet	180- 585	300- 720	675-1100	400-1500	200- 650	360-1500	230- 580	360-2100	210- 520	480-3350	1200-7200
	100 feet	180- 585	300- 720	675-1100	300-1100	0- 570	310-1200	210- 510	290-1530	180- 460	320-3120	120-6720
	125 feet	180- 585	300- 720	675-1100	220- 900	0- 500	270- 810	180- 450	270-1100	160- 400	160-2920	0-6300
	150 feet	180- 585	300- 720	675-1100	0- 750	0- 400	220- 670	170- 360	220- 670	150- 320	0-2640	0-5880
	200 feet	180- 585	300- 720	675-1100	0- 500	0- 300	180- 450	160- 270	180- 450	140- 240	0-2250	0-4320
	250 feet	180- 585	300- 720	675-1100	0- 330	0- 270	170- 290	150- 210	170- 290	130- 180	0-2000	0-3840
	300 feet	180- 585	300- 720	675-1100	0- 230	0- 180	160- 210	140- 160	160- 200	120- 140	0-1560	
	400 feet	180- 585	300- 720								0-1130	
	500 feet	180- 585	300- 720								0- 820	
	600 feet	180- 585	300- 720								0- 660	
	700 feet	180- 585	300- 720								0- 590	
	800 feet	180- 585	300- 720								0- 510	
	900 feet										0- 400	
	1000 feet										0- 270	
5" Diameter	30 feet	180- 825	300-1620	Not adaptable	Not adaptable	650-2950	1500-4075	650-2350	Not adaptable	1000-2000	640-4000**	2160-7860
	50 feet	180- 825	300-1620			540-2300	1085-3375	690-1800		790-1600	570-3600	1560-7560
	70 feet	180- 825	300-1620			450-1800	870-2875	420-1600		595-1440	480-3350	1200-7200
	100 feet	180- 825	300-1620			300-1300	610-2060	282-1100		468-1040	320-3120	120-6720
	125 feet	180- 825	300-1620			275-1000	390-1580	250- 900		400- 800	160-2920	0-6300
	150 feet	180- 825	300-1620			190- 800	0-1340	192- 720		220- 640	0-2640	0-5880
	200 feet	180- 825	300-1620			0- 550	0- 750	0- 440		190- 440	0-2250	0-4320
	250 feet	180- 825	300-1620			0- 350	0- 340	0- 310		0- 280	0-2000	0-3840
	300 feet	180- 825	300-1620			0- 260		0- 230		0- 210	0-1560	
	400 feet	180- 825	300-1620								0-1130	
	500 feet	180- 825	300-1620								0- 820	
	600 feet	180- 825	300-1620								0- 660	
	700 feet	180- 825	300-1620								0- 590	
	800 feet	180- 825	300-1620								0- 510	
	900 feet										0- 400	
	1000 feet										0- 270	
6" Diameter (and larger)	30 feet	180-1290	300-2160	Not adaptable	Not adaptable	930-3400	Not adaptable	2000-4000	Not adaptable	0-2700	640-4000**	2160-7860
	50 feet	180-1290	300-2160			930-3000		1800-3350		0-2400	570-3600	1560-7560
	70 feet	180-1290	300-2160			930-2000		1620-2600		0-1600	480-3350	1200-7200
	100 feet	180-1290	300-2160			680-1300		1080-1750		0-1040	320-3120	120-6720
	125 feet	180-1290	300-2160			590-1000		830-1200		0- 800	160-2920	0-6300
	150 feet	180-1290	300-2160			230- 800		0- 720		0- 640	0-2640	0-5880
	200 feet	180-1290	300-2160			0- 550		0- 500		0- 440	0-2250	0-4320
	250 feet	180-1290	300-2160			0- 350		0- 310		0- 280	0-2000	0-3840
	300 feet	180-1290	300-2160			0- 260		0- 230		0- 210	0-1560	
	400 feet	180-1290	300-2160								0-1130	
	500 feet	180-1290	300-2160								0- 820	
	600 feet	180-1290	300-2160								0- 660	
	700 feet	180-1290	300-2160								0- 590	
	800 feet	180-1290	300-2160								0- 510	
	900 feet										0- 400	
	1000 feet										0- 270	

*Minimum capacity based on 6" stroke at 50 strokes per minute. Maximum based on 9" stroke at 45 strokes per minute.
**Larger units can be built to fit individual needs.

PRESSURE RANGE NEEDED FOR ADEQUATE WATER SERVICE

The matter of what pressure range to use to provide adequate water service has become an increasingly important factor in selecting a pump. A **20 to 40 pound (psi) pressure range** has been considered standard by the pump industry for many years. That is, when the pressure drops to 20 pounds (psi), the pump starts automatically. It continues pumping until the pressure reaches 40 pounds (psi) and then stops automatically. As water is used the pressure lowers. The pump starts again when the pressure lowers to 20 pounds (psi).

Some thought is now being given to a **higher pressure range—possibly 30-50 psi or 40-60 psi**—as a result of recent user studies.[35] There are several reasons: (1) most automatic clotheswashers and dishwashers, and most water-conditioning units, work better at a higher pressure, (2) for hose cleaning higher pressure does a better job of loosening and removing waste materials, (3) some plumbing systems with undersize or partially clogged pipes need higher pressure to help improve a low flow rate, and (4) higher pressure provides better water delivery for fire protection.

Many *automatic water-using appliances* have a time-fill flow regulator that controls the water level in the unit. They commonly require a minimum pressure of 15 to 20 pounds (psi) to deliver an adequate supply of water. With a 20-40 pound pressure-range setting on the pump and a long run of pipe for the water to flow through to the appliance, the pressure is often as low as 5 to 7 pounds at the appliance inlet. At that pressure the flow regulator will shut off the water supply before it has reached the proper level in the appliance.

Some *water-conditioning equipment* causes a loss of from 15 to 20 pounds of pressure as the water flows through the unit. This means a very slow water flow at faucets served by a conditioner if there is a 20-40 pound pressure-range setting on the pump.

Studies conducted by the USDA show that, under most conditions, there is adequate cleaning action in *removing waste material* with a pressure-range setting of 35 to 55 pounds (psi) pressure. This is especially true if you can allow time for a short soaking period such as where there are rather limited areas to clean or where cleaning is done irregularly. For a pressure-range setting of 35 to 55 pounds (psi) to be satisfactory, the pipe size between the pump and the hose outlet must be large enough to keep down friction loss. Otherwise, considerable pressure is lost in overcoming pipe friction.

Where **labor and time saving** is important, a **higher water pressure—about 80 pounds** (psi)—is needed. This higher pressure is desirable for cleaning automobiles, tractors and field equipment as well as for removing animal waste from floors. It provides the necessary

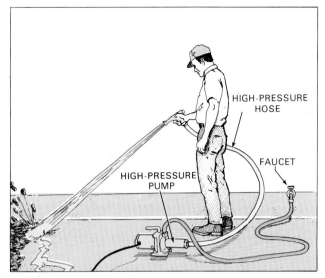

FIGURE 77. High-pressure portable pump for use in supplying pressures of 80 psi or more for cleaning purposes.

softening, soaking and loosening action, along with enough water to make the waste removal rapid and complete.[19]

Pressures substantially **higher than 80 pounds** (psi) develop needlepoint spray streams and misting. This frequently causes damage to painted surfaces and may pit concrete surfaces. At the same time it does not always provide enough water to adequately flush away loosened materials.

For these higher pressures, rather than increase the pressure range on your entire plumbing system, it is better to use a **specially-designed high-pressure pump.** For limited use, this can be a portable pump (Figure 77), which connects to your home water supply. It will add 80 pounds pressure or more to that supplied to the intake connection of the pump by your water-system pump.

Where **substantial quantities of water under high pressure** are needed for waste removal, a permanent installation is recommended. It would consist of a high-pressure pump, a specially-constructed high-pressure storage tank and separate plumbing.

If you have a **plumbing system already installed and the water flow rate is too low,** raising the pressure range on the pump will provide some limited relief. It will help make available some additional water but it seldom corrects the problem. It will not take the place of using larger piping sized to meet your needs. This point will be discussed further under "How to Select the Proper Supply-Pipe Size," page 131.

If you are interested in your pump developing enough **pressure for improved fire protection,** it should be able to maintain about 35 to 40 pounds pressure at the recommended fire pumping capacity of 10 gallons per minute.

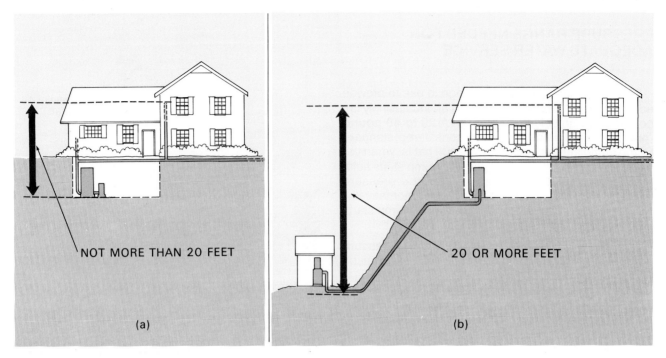

FIGURE 78. (a) Water outlets not more than 20 feet above pump level will have satisfactory pressure if proper size pipe is used between the pump and the outlet. (b) If your pump is more than 20 feet below the highest outlet, higher pump pressures will be needed.

HEIGHT WATER IS LIFTED ABOVE PUMP

If the **highest water outlets** supplied by your pump are **not more than 20 feet above the level of your pump,** you have no problem (Figure 78a). Here is the reason. As explained in the preceding discussion, the lowest operating pressure range for most water systems is between 20 and 40 pounds (psi). That is, at 20 pounds pressure the pump automatically starts pumping. It continues until the pressure reaches 40 pounds and then stops automatically.

Twenty pounds will supply fairly satisfactory pressure to your faucets, if your water outlets are not more than 20 feet above the level of your pump and if you use the right size of pipe to deliver the water with very little friction loss. Pipe size will be discussed later under "How to Select the Proper Supply-Pipe Size."

If you have a **shallow-well pump,** and the level of your pump is **more than about 20 feet below the level of your highest water outlet** (Figure 78b), this factor must be given special consideration. If you do not consider it, you will not get satisfactory water service at your highest faucets. In fact, if your buildings are as much as 100 feet above the level of the pump, you will not get any water. The reason—it takes one pound of pressure to push water to a height of 2.3 feet, so a height of 100 feet would require a pressure of 43.5 pounds just to raise water to that level. If the maximum operating pressure is only 40 pounds (psi), provision must be made for more pressure.

Pumping water to these higher levels may require a **heavier pump-and-motor combination.** This is especially true of shallow-well pumps (Figure 79). Table XI shows the types and capacities of shallow-well pumps for pumping water to higher elevations.

The same problem has to be considered if a **deep-well pump** is being selected to pump water to an elevation higher than 20 feet above pump level. In this case, the effect on the pump is the same as it is if the well is

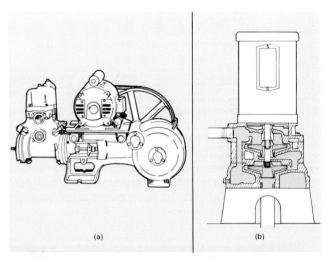

FIGURE 79. Heavy-duty shallow-well pumps used where extra pressure is needed to raise water to faucets located higher than 20 feet above level of pump. (a) Piston pump designed for high pressures. (b) Centrifugal pump equipped with an extra impeller to deliver water at higher pressure.

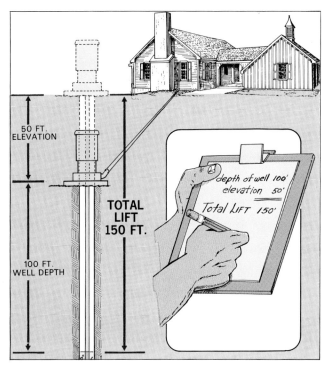

FIGURE 80. Where water must be raised more than 20 feet above the level of a deep-well pump, add extra elevation to depth of well (except with jet pumps) to get total lift in feet. Note Table XII.

deeper by the amount of the elevation. Note in Figure 80 the depth to water below the pump is 100 feet. The height the water must be lifted is 50 feet. If you add the two figures together, the total lift is 150 feet. That height is equal to about 65 pounds (psi) pressure. This is the total lift the pump must be designed to meet in addition to the pressure range you selected to provide adequate water service.

The *"total lift"* figures in Table XII can be used in this manner for all pumps except jet pumps. The table shows their capacities for 50- and 75-foot elevations above pump level. Although jet pumps are available for much greater depths, home- and farm-size jet pumps are not very efficient for lifting water to much greater heights.

PUMP LOCATION

A factor that may be of importance in the type of pump you select is whether or not it can be **offset from the well or water source.** The term *"offset"* means provision to place the pump and/or pressure tank from a few feet to a hundred or more feet away from the water source (Figure 81a and b), or to place the pump in the well and locate the pressure tank away from the well.

If you can use an offset installation, it may save you the **cost of building a pump house.** On the other hand, it is important that the place you select for your pump and tank, or tank only, be sanitary, protected from mechanical and weather damage and accessible for maintenance, etc., as outlined in "F. What Housing to Provide for Pump and Water Storage on page 116."

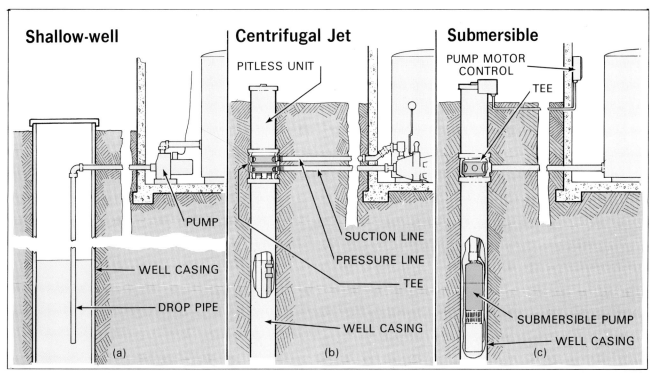

FIGURE 81. (a) An offset installation for a shallow-well pump. (b) An offset installation for a jet pump. (c) With a submersible pump, the storage tank is often offset from the well. This arrangement is also possible with other deep-well pumps.

95

All shallow-well pumps can be offset (Figure 81a) regardless of type if

1. *Ample pipe size* is provided between the water source and pump. This is important. Small pipes can cause so much resistance (friction) to water flow that the suction capacity of your pump is partly used up in overcoming the friction. The result—it cannot lift water from as great a depth.
2. *Total height* between water source and pump is still *within suction limits of pump.*

   In locating your pump away from the well, you may move to a higher location. The extra height must be added to the total suction lift. Make sure it is not more than the total suction lift of your pump.

Your dealer has recommendations from his manufacturer as to the proper size pipe(s) to use with your pump when set at various distances from the water source.

The **deep-well jet pump** can be offset as shown in Figure 81b.

With a **submersible pump,** only the tank is offset (Figure 81c). The pump is in the well rather than located in the basement, as shown in Figure 81a and b.

PUMP DURABILITY AND EFFICIENCY

According to durability studies, a water system should last an average of about 11 years. Some may last as long as 25 years.

Most users who have difficulty getting good service from their pumps have ones that do not meet their pumping needs, or else the pumps were not installed properly. For example, suppose you purchase a pump and find you have a weak water flow at some of your higher outlets. You then have the pressure switch adjusted to provide higher pressure.

If you have a positive-acting pump, the extra pressure may greatly overload the motor, causing it to burn out—unless the motor has ample overload protection. Even with overload protection, the motor may give only intermittent pumping service because the overload-protection device may turn the motor off when it gets too hot and turn it on again when it cools.

Suppose you purchase a centrifugal pump for the conditions just mentioned—one designed for low pressure but pumping against high pressure. It will either deliver very little water and run most of the time, or not deliver any water and run all of the time. In either case, pump wear is greatly increased over what it would be if the proper pump had been installed in the first place.

Pump efficiency is a matter of importance. Efficiency affects you most in the cost of operation. With electricity costing 7¢ per kWh and more, you should get the most efficient motor and pump.

Efficiency must be considered when a jet pump is used to pump large quantities of water from depths greater than about 100 to 125 feet. Overall pump efficiency decreases rapidly at greater depths. The greater inefficiency requires an increased motor size and possibly longer running time which, in turn, increases operating cost. However, you may still be justified in buying a jet pump for deep-well pumping if you can save installation cost by offsetting your pump as compared to providing a special structure for protection—or if your electric rates are such that overall cost is still reasonable.

DEALER SERVICE

Of the various factors you have to consider in buying a new pump, **dealer service** is one of the most important. A good dealer can perform two important services for you. He can

1. Help select a pump that will best fit your needs.
2. Supply emergency parts and service for your pump when you need them.

When **water service is interrupted,** it is always a serious inconvenience about the home. If you are using your water system for watering livestock and poultry, milk cooling, washing of dairy utensils, egg cleaning, etc., good service can easily pay for itself by avoiding any serious interruption in production. It can also pay good returns through emergency fire protection.

D. What Type and Size Water Storage to Use

If your water supply provides plenty of water for your need, and if you have selected the proper size pump, it will be easy to select the right size of pressure storage tank. The amount of stored water in the pressure tank is equal to the pump discharge in GPM.

In this section you will learn of **the different types of pressure storage tanks and water storage, how they work and the factors you need to consider in selecting the right one.** These are discussed under the following headings:
1. Types of Storages and How They Work.
2. Amount of Water to be Stored.
3. Sanitation Features of Different Storages.

1. TYPES OF STORAGES AND HOW THEY WORK

There are two general types of water storage facilities. They are discussed under the following headings:
a. Pressure Storage Tanks.
b. Gravity Feed and Pumped Storage.

a. Pressure Storage Tanks

Water is, for all practical purposes, incompressible. So it is necessary to provide a pressure tank with storable energy (air under pressure) which acts upon the water to force it to the point of use.

For many years, users of pressure tanks have faced the problem of the air supply in the tank being gradually dissolved by the water until very little air is left after a few months of use (Figure 82). This condition is called water logging. The result is the rapid starting and stopping of the pump which reduces its life expectancy.

Since another characteristic of water is its ability to absorb air, the design of the pressurized storage tank must have provisions to replenish the absorbed air or a means to prevent absorption.

Pressure storage tanks commonly used under these conditions are as follows (Figure 83):
(1) Precharged Diaphragm or Bladder Type.
(2) Plain Steel With Floating Wafer Type.
(3) Plain Steel.

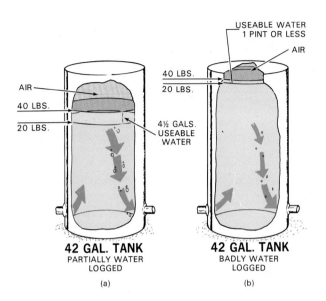

FIGURE 82. Water-logged pressure tank. (a) This is a condition where the air in the tank is gradually absorbed by the water until the tank can deliver only about 4¼ gallons of water of a 42-gallon tank as the pressure drops from 40 pounds to 20 pounds. (b) An extreme water-logged condition. Almost all of the air supply has been absorbed. Withdrawal of a pint or less of water from the tank reduces the pressure from 40 pounds fo 20 pounds pressure.

(1) PRECHARGED DIAPHRAGM/BLADDER TANK. The diaphragm or bladder type pressure storage tank shown in Figures 83a and 83b uses a mechanical locked-in flexible separator which completely isolates the air from the water. The tank is precharged at the factory. An air charging valve allows the installer to change the pressure if it's desired. An air volume control or air recharging system cannot be used on this type of tank since air and water are permanently separated. Only one water connection is required which serves as both inlet and outlet for the tank.

The diaphragm and bladder type designs are factory precharged to the low pressure switch setting (pump cut-in) allowing all the water that enters to be usable. This is the main reason why a diaphragm/bladder precharge design, in most cases, can be from 2 to 3.5 times smaller in total volume than the plain steel tank and deliver an equal amount of water.

Regardless of the tank design, as you use water, the compressed air in the tank pushes the water out through

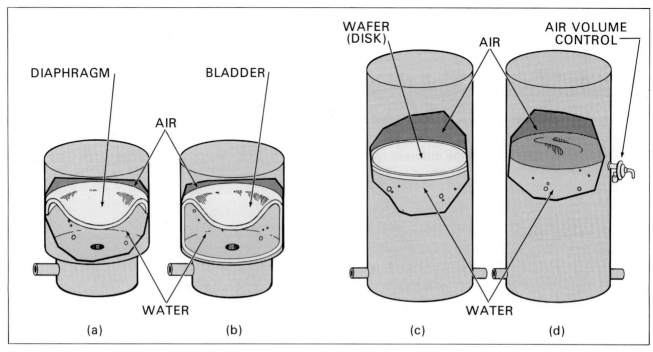

FIGURE 83. Types of pressure tanks. (a) Diaphragm, (b) Bladder, (c) Plain steel with floating disk, and (d) Plain steel.

the plumbing system. When the pump cut-in pressure has been reached, the pump will start automatically and begin to restore the water in the tank. As the water enters the tank, the captive air is compressed and the pressure increases until it reaches the cut-out (pump stops).

(2) PLAIN STEEL TANK WITH FLOATING WAFER TANK. The floating wafer type pressurized storage tank is illustrated in Figure 83c. It uses a wafer to reduce the air and water contact area. The wafer rides on the surface of the water. An air charging valve is used to replace the air. At installation, the tank is supercharged through the charging valve. Since the wafer doesn't completely separate the air from the water, subsequent supercharging is necessary. The water inlet and outlet are located near the bottom of the tank. This tank must be installed in the vertical position.

It should be noted that the plain steel tank with floating wafer air separation will deliver no additional draw-down unless it is supercharged after installation. After supercharging, the water level cannot drop to the outlet connection or air will escape under pressure to the fixtures.

(3) PLAIN STEEL TANK. The plain steel storage tank is shown in Figure 83d. This type of pressure tank can be designed for use in a vertical or horizontal position. The tank discharge should always be located near the bottom of the tank. This type has no barrier separating the air from the water. The tank has connections for an outlet and inlet near the bottom. There is also a separate connection for the air volume control usually located at the level where the water will be when the pump starts.

The main factor in the designs that affects the size pressure tank required is the initial pressure. As the initial tank pressure (precharge) increases, the draw-down increases up to a precharge pressure equal to the pump cut-in pressure. After this pressure has been reached, any further increase in the precharge pressure will result in a reduction in draw-down. Plain steel tanks are pressurized by pumping water into the tank. The water compresses the trapped air and increases its pressure. The water that is pumped into the tank to increase the pressure to the low pressure switch setting (pump cut-in) remains in the tank between pump cycles. The wafer type design is supercharged (pressurized with air at the job site) 5 psig below the low pressure switch setting in order to prevent the air from escaping. The water that is pumped into the tank between the 5 psig differential remains in the tank.

b. Gravity Feed and Pumped Storage

When the water source is of limited yield, you have to select a well pump with a discharge capacity that is somewhat less than the water yield from your well or provide a secondary water storage supply.

It must be sized so it is large enough to meet your peak demands and allow some reserve. Once the proper storage tank has been installed, the well system is now considered to have an adequate water supply and the pressure storage tank can be sized in relation to the pump capacity.

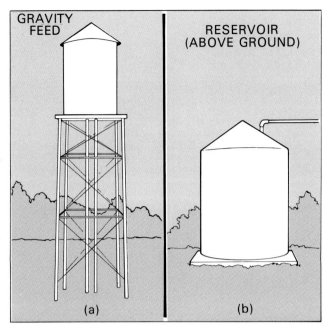

FIGURE 84. (a) Gravity tank. Depends on tank height to supply pressure. It can be used to store comparatively large quantities of water. (b) Reservoir. A surface or underground storage. It provides no pressure unless built about the level of the buildings it supplies. It is usually used to store large quantities of water.

There are two types of non-pressurized storage (Figure 84):
(1) Gravity-feed Tanks (Figure 84a).
(2) Pumped Storage (Figure 84b).

(1) GRAVITY FEED TANKS. A gravity tank is a large water storage located well above the level of the buildings. The higher the storage tank above your water-use outlets, the more pressure there is at your faucets. Water develops a pound of pressure for each 2.3 ft of elevation (height). To supply a minimum of 20 psig of pressure at your building, a tank would have to be mounted 46 feet (2.3 ft x 20 psig) higher than your faucets (Figure 85). As you can see from the illustration, the quantity of water being stored has nothing to do with pressure at your faucets. WATER LEVEL HEIGHT IS WHAT CAUSES PRESSURE.

In hilly country, the tank is usually placed on a hill to get needed elevation. In level areas, it may be located in an attic or on a tower built especially for tank support (Figure 84a).

In most areas, it is difficult to get enough tank height to provide satisfactory pressure. But, if you are in an area where you can get ample tank height, and your water source provides a limited amount of water, a gravity tank is one method of supplying the extra water storage you need for peak demands.

(2) PUMPED STORAGE RESERVOIRS. In this discussion, reservoirs refer to those water storages built approximately at pump level. They may be constructed either above or below ground. They usually consist of either a poured-concrete structure (Figure 84b), or a very large steel tank (Figure 86).

A tank used in this manner supplies little or no pressure. It acts only as a storage reservoir or "intermediate storage."

This intermediate storage reservoir is widely used when the water source (well yield) provides a constant but limited flow of water—not enough to take care of peak demands when they occur. By selecting a pump with somewhat less capacity than the yield of the well, you can pump over a long period of time and store the water in the intermediate storage to meet the peak demands. This is the pump in Figure 86 labeled "well pump."

With the intermediate tank arrangement, a SECOND PUMP is used. Its job is to pump water from the intermediate storage and put it under pressure. It is the pump in Figure 86 labeled "pressure pump." It must also supply enough water to meet peak demands. It is the pump with the capacity you determined under "A. What Capacity Pump is Needed." Both pumps work automatically. The well pump is controlled by water-level sensors located in the intermediate storage tank. The pressure pump is controlled by a standard pressure switch.

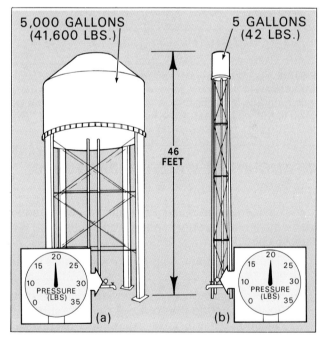

FIGURE 85. The pressure a gravity tank will provide is determined by how high the tank is above the level of your faucets. (a) A 5,000 gallon tank provides no more pressure than (b) a 5-gallon tank mounted at the same height.

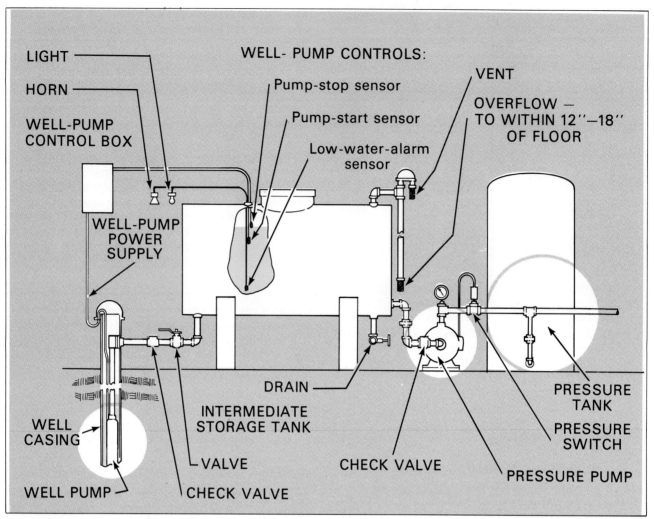

FIGURE 86. When a water source does not supply enough water to meet peak demands, some type of intermediate water storage is necessary. A low-capacity pump can be used in the well to match the water yield of the well. The pressure pump is then sized to meet the maximum water-use demand. The intermediate storage tank, if properly selected, supplies enough water for the pressure pump to meet all demands.

2. AMOUNT OF WATER TO BE STORED

In considering the amount of water you will need to store, there are two very important conditions you will need to consider:

(1) The rate of water yield from your present water source, and
(2) Whether or not you wish to provide a water reserve for fire-fighting purposes. A third condition is included in this discussion if you wish to consider it—the amount of safe water storage in case of a nuclear explosion.

Study the factors under the following headings:
a. Water Supply Yield.
b. Amount Needed for Fire Protection.
c. Amount Needed for Nuclear War Survival.

a. Water Supply Yield

If your well, or water source, provides plenty of water to meet your peak demands, and your pump is properly selected, the pressure storage tank can be selected from TABLE XIV.

The pressure storage tank sizing is based on: the pump capacity, design of pressure, and pressure switch setting. Up to a pump capacity of 10 GPM, the pressure tank drawdown is equal to the pump capacity, i.e. 5 GPM = 5 GALLON DRAWDOWN. This allows a minimum pump running time of one (1) minute. Beyond 10 GPM the drawdown becomes greater than the pump capacity in order to take into account the inertia of the higher horsepower pumps.

The pressurized storage tank sized in this manner provides two (2) major functions: (1) to protect and

TABLE XIV. PRESSURE TANK SELECTION CHART*

Pump, Capacity (Gallons per min.)	Pressure-switch Range Settings (Pounds per square inch)					
	20 to 40	30 to 50	40 to 60	50 to 70	60 to 80	70 to 100
	Pressure-tank Sizes (Gallons)					
4	42	82	82	120	120	120
8	82	120	180	220	315	315
12	120	180	220	315	315	315
15	144	220	315	525	525	525
18	180	315	315	525	525	525
24	220	315	525	525	1000	1000
32	315	525	525	1000	1000	1000

*Tank sizes indicated are ones that are commercially available.

prolong the life of the pump by preventing rapid cycling of the pump motor. (Many motor manufacturers recommend pump cycle rates of under 300 for each 24-hour period and not more than 30 starts per hour). And (2) to provide water under pressure for delivery when the pump is not running. Figure 87 depicts the proper installation relationship for the pump, pressure tank, and pressure switch.

One method of arriving at the size of the intermediate storage tank **assures enough water storage** for the pressure pump to operate continuously for two hours at full peak demand. This assumes the tank is full at the start of the two-hour period and that the well pump is working during this period. For normal use the peak demand would be for a much shorter period. Consequently, the storage tank actually provides a **substantial reserve of water** beyond normal needs.

It also provides enough reserve for considerable **protection in case of a fire.**

The USDA recommends that fresh-water storages of this type be no smaller than 2,000 gallons. There is little cost savings in smaller sizes.

To **determine the size intermediate storage tank** you need, proceed as follows:

1. *Write down the capacity of the pressure pump.*

    This is the capacity you determined under "A. What Capacity Pump is Needed." It is the pump that will draw directly from the intermediate storage tank and put the water under pressure to the various water-use outlets. Consequently, it is called a "pressure pump" in this part of the discussion.

    Example: Suppose you determined the pressure pump needs to have a capacity of 25 gpm.

2. *Determine the difference in pumping capacities between the well pump and the pressure pump.*

    The well pump is the one lifting water from the well into the intermediate storage tank. It should have a pumping capacity at least 10% less than the water yield you measured under the heading "Determining the Amount of Water Available."

    Example: Suppose that your well pump has a pumping capacity of 5 gallons per minute.

    Figure the capacity difference as follows:

Pressure-pump capacity	25 gal./min.
Well pump	5 gal./min.
Difference in pumping capacity	20 gal./min.

3. *Determine the size tank needed for intermediate storage.*

    The size tank needed is usually based on a 2-hour period of heavy continuous use.

    Example: for a 2-hour period, figure as follows:

    20 gal./min. × 120 min. (2 hrs.) = 2,400 gals. tank size.

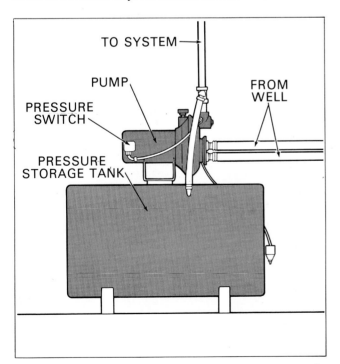

FIGURE 87. Proper installation for pressure storage tank, switch and pump.

b. Amount Needed for Fire Protection

If you are interested in a reserve water supply for fire or emergency use, Figure 88 shows how this can be done. A second storage is provided under the floor of the water-system center. With a reserve storage capacity of 5,000 gallons or more, and a fire-hydrant connection attached to it, a substantial quantity of water is available for **community fire-fighting equipment.** Studies conducted by the Agricultural Research Service show that a water system center of this type, if centrally located in relation to the buildings it supplies, is an excellent answer to providing ample water for all needs from either a limited or adequate water supply.

In Figure 88 the well is shown next to the water-system structure. This is not necessary. **In placing the water-system center** where it will be well located in relation to all of the buildings, it **may be a considerable distance from the well.** Since both storages collect water over a period of time, the well pump can be of relatively low capacity. This in turn, means that the well may be several hundred feet from the center, and a relatively small water line can be used between the well and the center. Also, the pump control circuit between the center and the well pump is not difficult to install.

The center makes possible **several supply lines**—even one to each building—which helps keep down pressure loss and makes it convenient to turn off the water supply to one building without affecting the others.

c. Amount Needed for Nuclear War Survival

If you wish to take into account the amount of water storage needed in case of a nuclear attack, Government recommendations are as follows:

> Since the water supply systems probably will not be operative after an attack, a 14-day supply of water for each member of the family should be stored. People can survive on ½ gallon per day for drinking and food preparation, but it would be far better to have at least 1 gallon per day per person. . . .
>
> Normally safe water that comes from properly sealed and covered wells will not generally be affected by emergency conditions; however, some means of getting water from the well should be available in the event the electric service is disrupted.

You can easily arrive at how much storage you will need for 14 days by checking the amount of water you have already determined you will need daily, page 12.

As shown in Figure 44, radioactive fallout consists of **radioactive dust.** When the dust is sealed out of the water supply, the water can continue to be used if there is some way of making it available. Some people take the precaution of securing an engine-driven generator that can be used to drive the automatic pump, or they install a hand-operated pump.

In case you have taken no precautions, the water in your pressure tank, water heater, and plumbing system is well protected. It may be enough to supply your family needs for as long as 14 days.

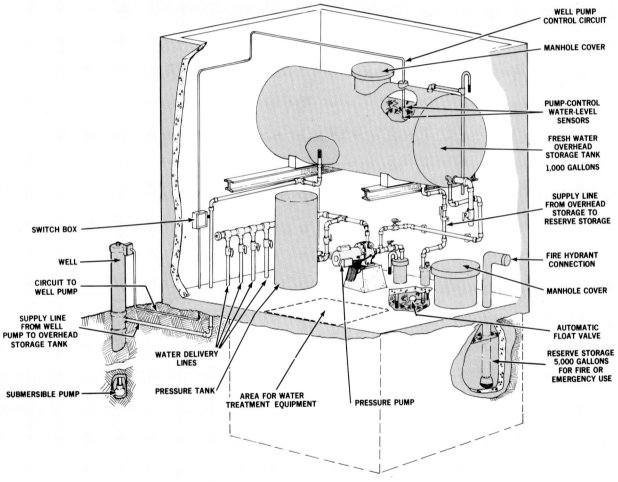

FIGURE 88. A water-system center with two storages. The overhead (Intermediate) storage is for day-to-day use. The underground reserve water storage is for two purposes: providing ample water for community fire equipment and for emergency use in case the overhead storage supply is not adequate. The building can be sized to house water-treatment equipment. Inset-Exterior view of water-system center. (USDA Plan No. EX5963, Farmstead Water Supply Center. Available through any State Cooperative Extension Service).

3. SANITATION FEATURES OF DIFFERENT STORAGES

Both the **elastic pressure cell** and the **pressure tank** are considered the most sanitary of the various storage methods. Since these units are completely enclosed, there is no way for them to become contaminated except as contaminated water is pumped into them or when pump connections are broken for servicing.

A **covered gravity tank** mounted on a hill or on a tower is difficult to completely seal at the top. As a result, dust, insects, rodents, and even birds may get into the tank and cause a pollution problem. An **open gravity tank** should not be considered. It is certain to become polluted.

A reservoir mounted above ground (Figure 89a) can be built tightly enough to keep pollution to a minimum. If the reservoir is built below ground (Figure 89b), there can be an additional problem of keeping out ground water if a crack should develop in the floor or walls, but that is not usually a problem if the tank is well built.

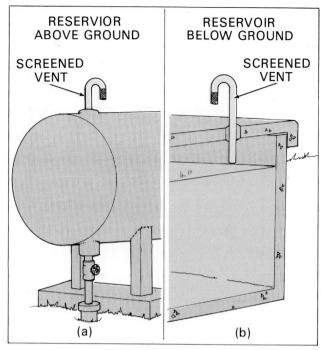

FIGURE 89. A screened vent is necessary to keep out insects and foreign matter with tightly-built reservoirs. The vent allows air to move in and out of the tank as water is drawn out or is pumped in.

E. Understanding Water-System Control Units

Thus far, your job has been to select a pump of the right pumping capacity and the right type to meet your needs. You then selected a water storage unit to match your pump and water reserve needs. Now you need to understand the controls your dealer will supply as part of the system or will recommend that you install in addition to those normally supplied.

Most water systems are sold with the more important accessories included. However, there is a wide variety of conditions to which pumping systems must be fitted. Your dealer may have occasion to change some of the accessories to meet your conditions. The more important ones are discussed here so you will have an understanding of what they are and how they work.

The various conditions your dealer must consider and the control units used to meet these conditions are discussed under the following headings:
— Water-supply control switches
— Pressure-tank air-volume controls
— Pump-prime controls
— Pressure-relief valves
— Pressure gages
— Water-hammer controls
— Sand-removal units

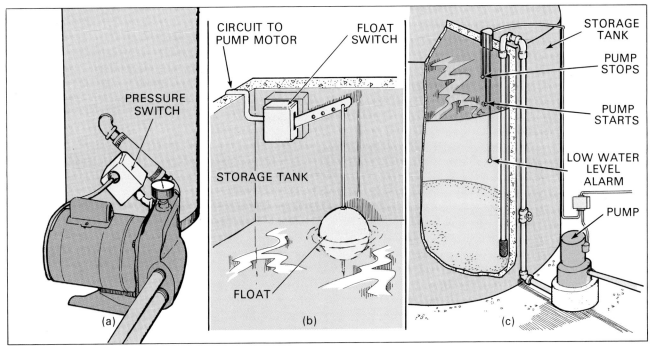

FIGURE 90. Types of pump-control units. (a) A pressure switch which responds to the pressures being developed within the water system by the pump. (b) Float switch for controlling the water level in gravity tanks or reservoirs. (c) Drawdown-control switch with water-level sensors for controlling the pump.

WATER-SUPPLY CONTROL SWITCHES

There are three types of **water-supply control switches** for control of the motor on your water pump. They are:

1. Pressure switches
2. Float switches
3. Drawdown-control switches

If you are planning to use an *elastic pressure cell or a pressure tank,* your dealer will provide a pressure switch (Figure 90a) to control pump operation.

If you have a *gravity tank or a reservoir,* your dealer will supply either a float switch (Figure 90b) or a drawdown-control switch (Figure 90c).

Of these three, **pressure switches** are by far the most common since they control pumps which develop the necessary pressure for water distribution. There are several different designs, but they all work on the general principle of the one shown in Figure 91. A rubber diaphragm in the pressure switch is exposed to pressure on the delivery side of the pump. The pressure switch may be on the pump, as in Figure 90a, on the pipeline between the pump and tank, or on the tank.

As the *pump builds up pressure,* pressure increases on the switch diaphragm. The diaphragm moves up and gradually compresses the large range spring. This action moves the lever that operates the cam wheel. As it is forced over the top of the operating cam lever,

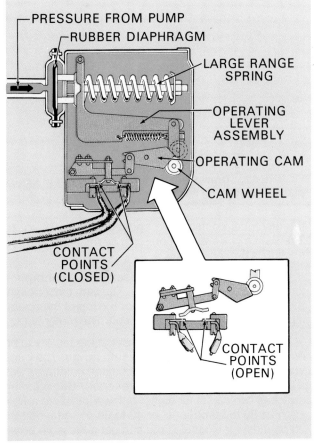

FIGURE 91. One type of pressure switch used with pressure cells and pressure tanks.

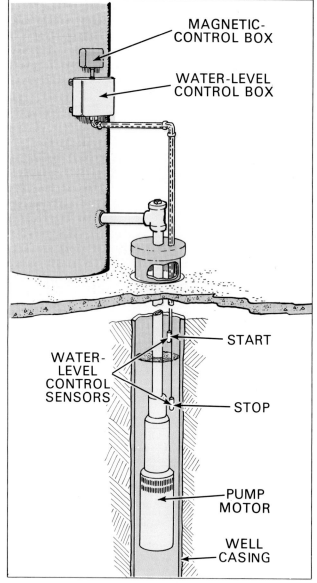

FIGURE 92. Drawdown-control switch used to control pump on wells with limited water yield.

If you are using a **gravity tank, or reservoir, a float switch** is used to control the pump motor (Figure 90b). The switch mechanism is similar to that of a pressure switch. But instead of being activated by water pressure, it is controlled by the float that follows the water level up and down in the tank or reservoir.

The only way a *float switch helps maintain pressure* is by keeping the water level in a gravity tank well towards the top of the tank. This helps keep the pressure in the plumbing system at a maximum during most of the time that water is being used. When the water level lowers, the float automatically trips the switch and starts the pump. When the water level rises, the float moves with it and turns off the pump motor when the desired water level is reached.

The **drawdown-control switch** can be used in place of a float switch in a *gravity tank or reservoir.* It consists of electric sensors (or probes), which can be adjusted to the desired positions in a water storage so as to start the pump motor when the water level gets low and stop it when the water reaches the desired height (Figure 92).

A *drawdown-control switch* can also be used where there is a *limited water yield* from a well. The two sensors are lowered down into the well. One sensor is placed at the lowest drawdown level desired, the other at a level where the pump should start when the water rises to that point (Figure 92). When the pressure switch or float switch starts the motor, as the water level drops to the level of the lower sensor, the pump motor is stopped even though the pressure switch, or float switch, has not turned off. The pump motor will not start again until the water level rises high enough to touch the top sensor. Other types are available.

the electrical contacts are opened (Figure 91, inset). The switch is a snap-action type, so the electrical connection is broken quickly.

When *tank pressure decreases,* the diaphragm is forced down by the range spring. The cam that operated the switch mechanism rides over the top of the operating cam lever and closes the circuit. The motor starts.

Most pressures switches are set at the factory to operate on a **20 to 40 pound pressure range.** The pump starts operating at 20 pounds (psi) and continues until a 40-pound (psi) pressure is reached. But pressure switches can be set for higher or lower pressure ranges such as 30 to 50 pounds, or 40 to 60 pounds, or 15 to 35 pounds, for example. You can get a pressure switch that will also cut the pump off when the pressure drops below 10 psi.

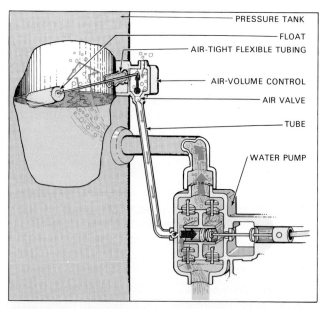

FIGURE 93. Shallow-well float-type air control. When the water level in the pressure tank gets too high, the float opens the air valve in the control. Air is drawn by the pump through the control into the pump and forced into the tank along with the water.

PRESSURE-TANK AIR-VOLUME CONTROLS

If you have a pressure tank, maintaining an air supply is very important as was explained under "What Type and Size Water Storage to Use." An air supply is not necessary for any of the other storage units.

If your **pressure tank is equipped with an absorption-barrier float,** there may be so little air absorbed by the water that adding air about every 9 to 12 months with a hand pump or air compressor is all that is needed.

If your **tank has no absorption-barrier float** or loses air rapidly even with the float, you will need an *air-volume control.* The control is used to keep the proper amount of air in the tank in relation to the amount of water. The controls are of several different designs to match the various type of pumps. In general, they are designed either (a) for piston pumps or (b) for centrifugal jet and turbine pumps.

PISTON PUMPS—If you have selected a **shallow-well piston pump** with a pressure tank, you will need a *shallow-well float-type control.*

Figure 93 shows how the shallow-well float-type control works. When the **water level in the pressure tank gets too high,** the float riding on the water surface inside of the tank is high enough to open the air valve in the control. This allows the pump to draw outside air into the air-volume control, past the air valve and into the tube that connects to the suction side of the pump. From there it is drawn into the pump where it mixes with the water and moves with the water into the pressure tank. The air then separates from the water and collects in the top of the tank.

As *air gradually accumulates in the tank,* the water level lowers. The float rides at a lower position. Finally a point is reached where the float will not open the air valve at any time during the pumping period. This shuts off air to the pump until the water level in the tank rises high enough to again open the valve.

If the *water level from the water source* is almost on a level with the pump, or a little above it, there will not be enough pump suction developed to pull the air through the control. This is a common situation when a pump is placed in a basement to pump water from an adjoining cistern or reservoir. In that case, a valve has to be placed in the suction line to restrict water flow enough to cause suction.

If you have selected a **deep-well piston pump** with a pressure tank, you will probably use an **air-pump air-release type of control.** With this type of control, instead of air being supplied to the tank only when needed, air is supplied continuously while the pump is operating (Figure 94 lower inset).

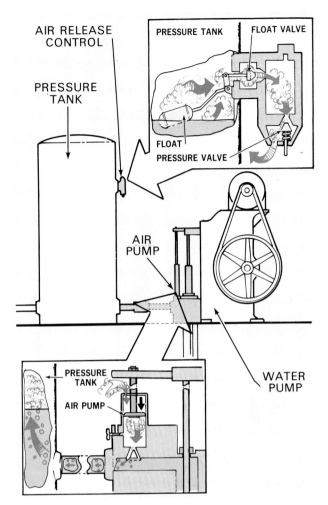

FIGURE 94. (Lower inset) With an air-pump air-release control combination, the air pump on the water pump delivers air continuously to the pressure tank. **(Upper inset)** The control in the tank simply releases excess air when the air volume becomes too great.

When the *water level is high* and more air is needed, the float rides high in the tank and keeps the float valve closed. This allows the incoming air to remain in the tank.

When *too much air accumulates* in the tank, the float rides low enough to keep the float valve open (Figure 94, top inset). Air can pass from the pressure tank into the outer chamber of the control and escape.

A pressure valve in the outer chamber keeps too much air from escaping when the tank pressure is below 30 pounds with a pressure-switch setting of 20-40.

CENTRIFUGAL, JET AND TURBINE PUMPS—If you have a *shallow-well turbine pump or certain of the centrifugal and centrifugal-jet pumps,* which are mounted in a horizontal position, you can use a **shallow-well float-type control** shown in Figure 93. This is due to their ability to pump some air without losing prime. But

107

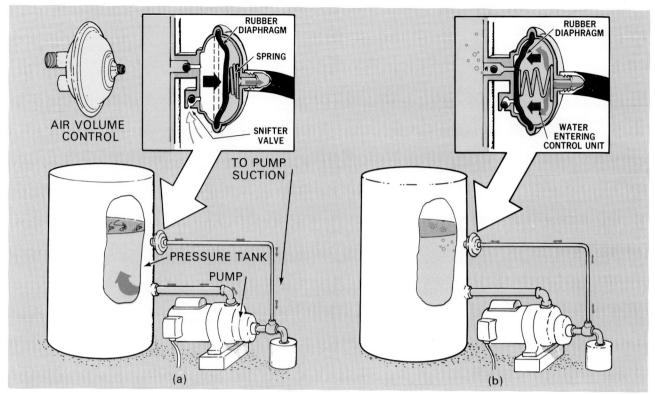

FIGURE 95. Diaphragm type of air control. (a) When pump starts, diaphragm is drawn to right by pump suction. Air enters through snifter valve, if water level in tank is above connection to control. (b) When pump stops, pressure equalizes, diaphragm moves to left by spring pressure and forces air into tank.

their pumping efficiency is greatly reduced. So they are generally equipped with one of the following:

1. Diaphragm-type air control
2. Water-displacement control
3. Venturi-and-air-release control

The purpose of these two controls is to **provide air** for the pressure tank **without having any of it pass through the pump.**

Figure 95 shows the **diaphragm-type air control.** Here is how it works.

Note in Figure 95a *as the pump starts, water is removed from the right side of the control* by pump suction (Figure 95a inset) causing a partial vacuum. This causes the diaphragm to move to the right against a spring. The sudden action causes a partial vacuum to develop in the control on the left side of the diaphragm. If the water in the pressure tank is above the opening connecting the tank to the diaphragm, the rush of water through the opening causes the ball valve to close. At the same time the partial vacuum causes outside air to force open the snifter valve, and a charge of air enters the control.

When the pump stops, all suction is removed from the diaphragm (Figure 95b), the spring forces the diaphragm to the left. This action forces the *new charge of air* through the small opening into the tank.

If the *water level in the tank is below the level of the control* when the pump starts, air moves from the tank into the control instead of through the snifter valve. When the pump stops, the air charge is forced back into the tank—no new air is added. With this arrangement the tank is supplied with new air only as needed.

This type of control *must be matched with the size of the tank.* The pump must also turn on and off frequently, if it is to supply enough air, since it supplies air only at the end of each pumping cycle. Also, a strong suction is necessary for this type of control to operate properly.

The principle involved in a **water-displacement type** of air control can be used in different ways. Figure 96 shows one method when connected to a pump suction line. A strong suction is not needed to make it work. In fact, it will work if there is a slight pressure in place of suction in the suction line—this is sometimes the situation with a deep-well jet pump.

When the pump starts, there is an immediate difference in pressure between the point where the control connects to the pressure tank, and the point where the suction tube connects to the pump suction line. The highest pressure is at the tank connection. If *water in the tank is above the opening to the control,* water from the tank passes through the jet into the air-control cylinder. The result—suction develops at the snifter valve

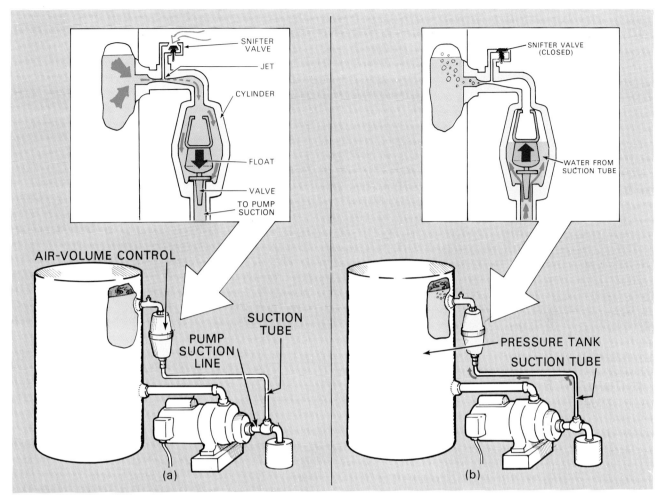

FIGURE 96. Water-displacement type air-volume control. (a) When the pump starts, water is drawn by pump suction from the cylinder of the air charger. Water from the pressure tank passes through the jet, causing outside air to be drawn through the snifter valve into the air-volume control cylinder. (b) When the pump stops, water pressures equalize. Water moves from the pump suction line into the air-control cylinder, forcing the new air into the pressure tank.

—air is carried with the water into the air-control cylinder until the air-control float valve seats in the charger. This stops further water removal from the cylinder.

When the pump stops, the water pressure equalizes —becomes the same in the pressure tank, in the pump and in the pump suction line. Water moves from the pump suction line back into the air-control cylinder. The air-control float valve is lifted, and the cylinder fills with water. The new charge of air from the air-control cylinder is forced into the tank.

If the *water level is below the tank connection to the control* when the pump starts, air flows into the cylinder from the tank. The air movement from the tank does not develop enough suction in the jet to draw in outside air. Consequently no new air is added to the tank supply.

This type of charger is *dependent on the pump starting and stopping frequently,* because a new charge of air is added only after each pump operation cycle.

Figure 97 shows a **second type of water-displacement air control.** With this unit, suction is developed from a jet placed in the supply line feeding water from the tank. When enough water is being used to develop a fairly substantial flow of water through the supply line, enough suction is developed by the jet to create a partial vacuum in the air-charger cylinder.

When the *pressure tank needs air,* the water level in the tank is above the level of the connection that leads to the air control. Water being removed from the air-charger cylinder causes a partial vacuum and water from the pressure tank starts flowing past the ball check valve into the cylinder. This water flow causes the ball

109

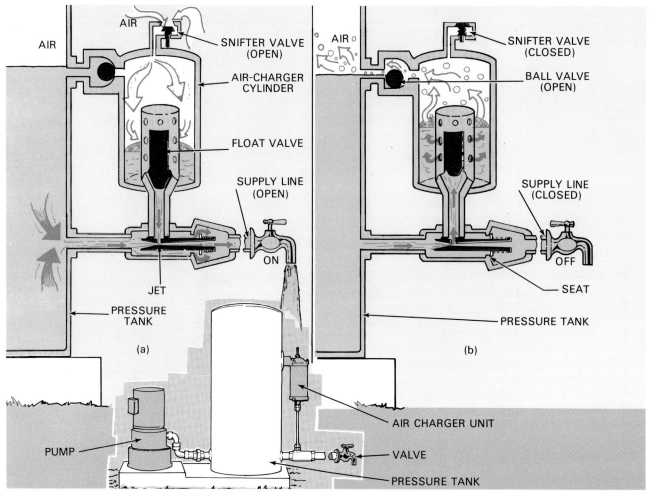

FIGURE 97. Water-displacement air charger with jet in supply line. (a) Movement of water through supply line from tank causes suction at jet. Suction draws water from air charger. Water from pressure tank closes ball-check valve. Partial vacuum in charger causes air to enter through snifter valve if tank needs air. (b) When water flow in the supply line stops and water pressure equalizes, water enters the air-charger cylinder and forces the air charge into the pressure tank.

check valve to close (Figure 97a). The partial vacuum then causes the snifter valve to open so air can enter the cylinder.

As the water continues to be drawn through the supply line, water continues to be removed from the air-charger cylinder by jet action until one of two conditions develop. Either the movement of water through the supply line is stopped, or the water level in the charger lowers until the float valve seats and stops further withdrawal from the charger.

When the *flow through the supply line is stopped* the water pressure equalizes in the supply line and the tank. This causes water from the supply line to force the float valve open and fill the charger cylinder. The ball check valve opens and the new charge of air flows into the pressure tank.

If the *pressure tank has plenty of air,* the water line is below the level of the control connection when the pump starts. Air moves from the pressure tank, past the ball valve and into the cylinder without the ball check valve closing. As a result, no outside air enters through the snifter valve.

When the flow through the supply line stops and the water pressure equalizes, air that was originally supplied from the pressure tank is returned to the tank.

When there is an *extra heavy flow of water* through the supply line, the venturi unit moves off of its seat to allow water to pass around it as well as through it.

One of the most simple and effective types of controls for submersible pumps is the **water-displacement type** shown in Figure 98. Here is how it works. When the *submersible pump has finished pumping,* the bleeder valve, located above the pump on the discharge pipe, opens and allows the water above it to drain out. The partial vacuum that develops, as the water bleeds from the pipe, causes the snifter valve to open next to the check valve so that air may enter to replace the drained water. The check valve stops any return water movement from the pressure tank. When all of the water is drained between the snifter valve and the bleeder

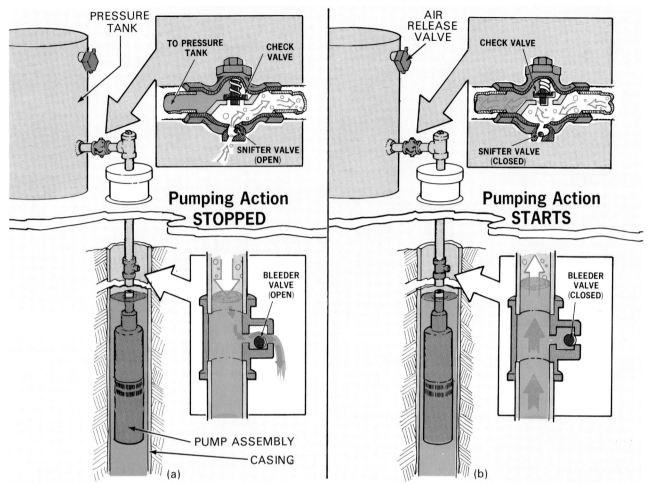

FIGURE 98. Water-displacement air-control design for submersible pumps. (a) When pumping action stops, the water between the bleeder valve and the check valve is drained through the bleeder valve. Air enters at snifter valve. (b) When the pump starts, air in the pipe is carried with the water to the storage tank. Excess air is released through the air-release valve in the pressure tank.

valve, a rather sizeable amount of air has entered the pipe.

As quickly as the *pump starts,* the pressure and water movement developed by the pump causes both the bleeder and snifter valves to close. The air charge trapped in the pipe is forced into the pressure tank along with the water.

The pipe above the bleeder valve is drained and recharged with air each time the pump stops. A new charge of air enters the tank each time the pump starts. When the *tank accumulates too much air,* it is released through the air-release control at the top of the pressure tank. This is the same control as shown in Figure 94.

Still another design is the **venturi-and-air release combination** (Figure 99). The venturi unit is placed between the pump and pressure tank. The pressure-drop adjuster valve is closed enough to force some water through the venturi as it passes from the pump to the pressure tank. Water passing through the venturi causes a partial vacuum at the snifter valve. The partial vacuum is enough for outside air to move through the snifter valve and into the water line where it is conveyed with the water to the pressure tank.

When *too much air accumulates* for proper tank operation, the air is released through an air-release type of control shown in Figure 94.

You can use the venturi and air-release air control on any of the centrifugal or turbine pumps.

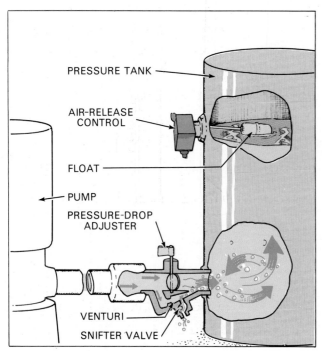

FIGURE 99. Venturi and air-release combination air control. A portion of the water passing from the pump to the pressure tank is diverted through a venturi. A partial vacuum develops—enough for air to enter through the snifter valve and pass into the pressure tank. Excess air is released by the air-release control at the top of the tank.

On some self-priming pumps a **check valve** (Figure 100b) is used on the suction line next to the pump. It is also used where two or more pumps are supplying water to the same delivery line to protect one pump from the pressure being developed by the other. Or, where water must be held in the pressure tank as in Figure 98.

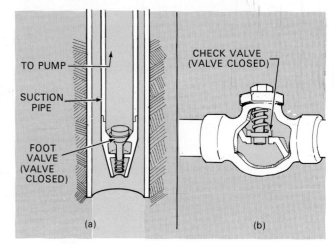

FIGURE 100. (a) Foot valve. Used on the end of a pump suction line to hold water in the suction line so the pump remains primed while not in operation. (b) Check valve. Permits water movement in only one direction. Sometimes used in suction line next to the pump.

PUMP-PRIME CONTROLS

Foot valves and check valves are usually supplied with your pump if they are considered necessary for maintaining pump prime. Both do the same kind of a job—permit water to move in only one direction.

A **foot valve** (Figure 100a) is installed on the end of the suction pipe in the well. It keeps the suction pipe below the pump full of water while there is no pumping action. This is especially important for all centrifugal and turbine-type pumps, since many of them pump little if any air, and their efficiency is greatly reduced when they do pump air.

Piston pumps can pump air, but if the valves are worn, or if there is a small leak in the suction line, a foot valve helps the pump to keep operating until the suction line can be repaired.

PRESSURE-RELIEF VALVES

If you have a **shallow-well or deep-well piston pump**, a relief valve (Figure 101, inset) is used to relieve excess pressure in case something happens so that the automatic pressure switch does not function. Instead of the pressure continuing to build to where the motor, pump or tank is damaged, the relief valve opens. Enough water escapes to keep the pressure within safe limits.

Centrifugal or jet pumps seldom have a relief valve. As pressure on these pumps increases, pumping capacity lowers. The pressure finally reaches a point high enough that all pumping action stops. That point is usually still low enough to prevent any damage to the tank and pump.

Note that the relief valve has a *pressure adjusting screw* so it can be adjusted to work at higher or lower pressures. It is important that an inexperienced person not tamper with this adjustment. If adjusted for extra high pressures, it may provide no protection for your pressure tank. A pressure tank explosion can cause considerable damage.

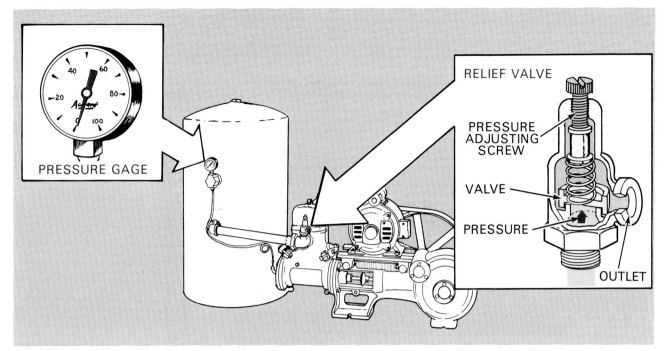

FIGURE 101. A relief valve is used to protect a water system from too much pressure. The pressure gage indicates storage-tank pressures. It is also used for adjusting the water pressure supplied to a deep-well jet on a jet pump.

PRESSURE GAGES

A **pressure gage** (Figure 101 inset) is standard equipment on most water systems. It is supplied to show the amount of pressure being developed by the pump. It is sometimes very helpful in locating pump trouble. If it is necessary to change the settings on your pressure switch, a pressure gage must be used.

On deep-well jet pumps, a pressure gage is located on the discharge side of the pump. With it, the serviceman can adjust the control valve on the pump discharge so as to return enough water at the proper pressure for the jet in the well to work satisfactorily.

WATER-HAMMER CONTROLS

Water hammer is usually recognized by the sudden thud and pipe vibration that occurs when a water valve closes suddenly. This often happens with automatic valves used on clothes washers and dishwashers. It also happens when water is being pumped up hill through a long pipe line.

The **cause of the water hammer** is the ramming action that develops when a column of water is suddenly set in motion as when a pump starts, or when the water column is suddenly stopped as when a valve is closed quickly.

If you are using a **pressure tank** or an **elastic storage cell,** these units usually absorb most of the water hammer in the immediate area where they are located.

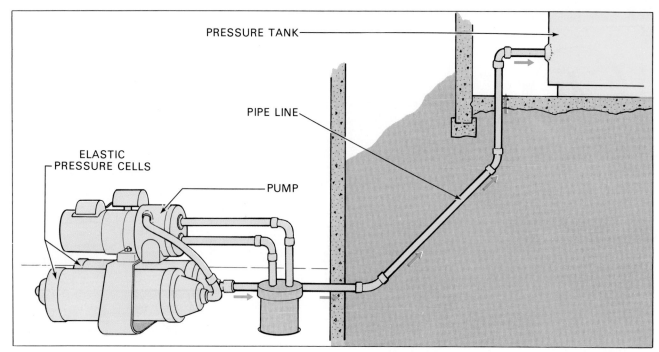

FIGURE 102. Means for controlling water hammer when water is pumped up hill. Elastic pressure cells located near the pump, and either a pressure tank or an elastic pressure cell at the upper level provide the necessary cushioning.

The biggest problem is where **water is being pumped up hill** through a considerable length of delivery pipe, and the pressure tank is located at the higher level where the water is being used (Figure 102).

The problem can be overcome by using **two pressure tanks**—one at the pump level and the second at the upper or delivery-point level. But this often creates air problems in both tanks. The air supply in the lower tank is under high pressure so the air is absorbed by the water rather readily and delivered to the upper tank. This makes it difficult to keep the lower tank properly charged with air.

A much better solution is to use **elastic cells next to the pump** to do away with the air problem. Then, you can use either an elastic pressure cell or a pressure tank at the upper level.

SAND-REMOVAL UNITS

If you **pump sand from your well** occasionally and are using a *shallow-well turbine pump*, a sediment trap may be a good investment. Sand will gradually cut away the finely-machined surfaces of the pumping mechanism until it loses much of its pumping capacity. To a lesser extent this is true of other types of pumps.

The **sediment trap** (Figure 103) is a steel container mounted in the pump suction line. Water from the water source entering the container at the top is deflected down by a baffle. Speed of water travel through the container is greatly reduced, allowing time for the sand to settle to the bottom of the container. Water, free of sand, then moves up the opposite side of the baffle and into the suction side of the pump. To be effective, a sediment trap must be cleaned frequently.

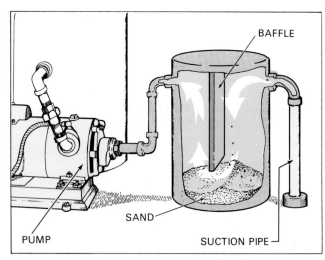

FIGURE 103. Sediment trap used with shallow-well turbine pumps. Water containing sand passes through tank at greatly reduced speed. Sand load is dropped to bottom of tank.

SAFETY CHECK VALVES

In many instances the farm well must meet all farm and home needs including demands for "fast fill" of large tanks and sprayers which encourage direct withdrawal from the well at rapid rates.

These fast flows can be easily obtained by tapping the main line at or near the well cap. However, **back siphoning** may occur without adequate **check and vacuum relief valves** located in the feeder line between the well and point of delivery, resulting in materials entering the well or domestic water supply.

The vacuum relief valve costs $1 to $10 and the check valve from $15 to $75 depending on size. Installed at the points shown in Figure 104, back flow can be prevented.

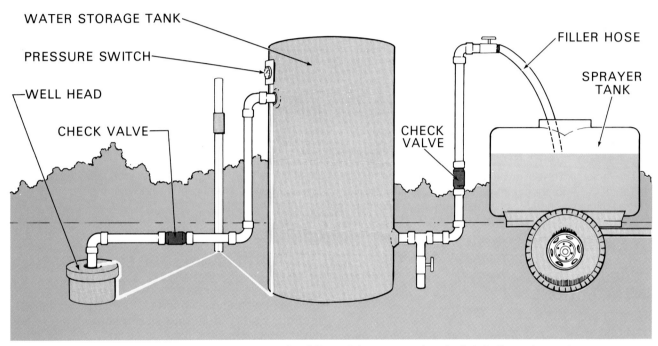

FIGURE 104. Location of check valves to prevent backflow of dangerous chemicals into the water system.

F. What Housing to Provide for Pump and Water Storage

Now that you have selected the pump you need and the type of water storage that fits your needs, your next decision is where and how to install them. If you are in an area where freezing is a problem, your major concern will probably be how to protect the system from freezing during cold weather. In some areas there is a problem of protecting the water system from heat, rain, or wind-blown dirt.

For you to make a satisfactory decision, you will need to understand the various types of water-system installations. Then consider the following factors in determining which type of installation will best fit your needs:
- Protection from surface water entering well
- Adequacy of weather protection
- Adequacy of drainage
- Adequacy of ventilation
- Ease of cleaning enclosure
- Ease of servicing equipment
- Ease of servicing well

The factors involved in your selection are summarized in Table XV. Additional explanations regarding certain of the factors you need to consider further are discussed under separate headings.

TYPES OF WATER-SYSTEM INSTALLATIONS

The basic types of water-system installations are shown in Figures 105 and 106.

ADEQUACY OF DRAINAGE

Drainage has been a serious problem with **pit-type installations** (Figure 106). Even when an adequate drain is installed, there is a problem of it becoming clogged. If a drain can be extended a short distance to a surface outlet, such as the side of a hill, it is less likely to become clogged. But, in many instances, it is difficult or impossible to provide that kind of drainage.

It is then that a **sump**—an underground opening filled with gravel or coarse material—is used for collecting excess water (Figure 106a). Supposedly, the water will drain into the adjoining soil. But, if the water table rises around the pit, the sump will fill with water and then feed it back into the pit. The water level in the pit rises until is it level with the water table outside of the pit.

Besides the possibility of **well pollution,** there is a strong possibility of the motor, the motor controls, or the **wiring system becoming shorted.** This will stop the pump. You may have considerable delay and expense getting your water system back into operating condition.

ADEQUACY OF VENTILATION

Ventilation is important because of **small leaks** that may develop around the pump and tank and cause water to accumulate.

In humid areas, **moisture condenses** on the pressure tank and drops to the floor.

These conditions keep parts of the floor damp or wet most of the time. Adequate ventilation, along with drainage, helps to prevent rusting and electrical trouble due to short circuiting.

EASE OF CLEANING THE ENCLOSURE

Ease of cleaning is important to keep down the **collections of dirt, dust and spiderwebs**. If the water-system enclosure is difficult to clean, it may never be cleaned.

Unclean conditions can cause (a) air snifter valves to clog, (b) motor burnouts due to clogged ventilation passages in the motor, and (c) entry of dirt into the water-delivery system during pump servicing.

EASE OF SERVICING THE EQUIPMENT

If you need to **replace your pump or your pressure tank,** the new unit may be of a different dimension than the unit you are replacing. Consequently, in determining the size of your enclosure, it is important that reserve space be provided.

It will also help your serviceman to do a better and faster job of servicing your equipment.

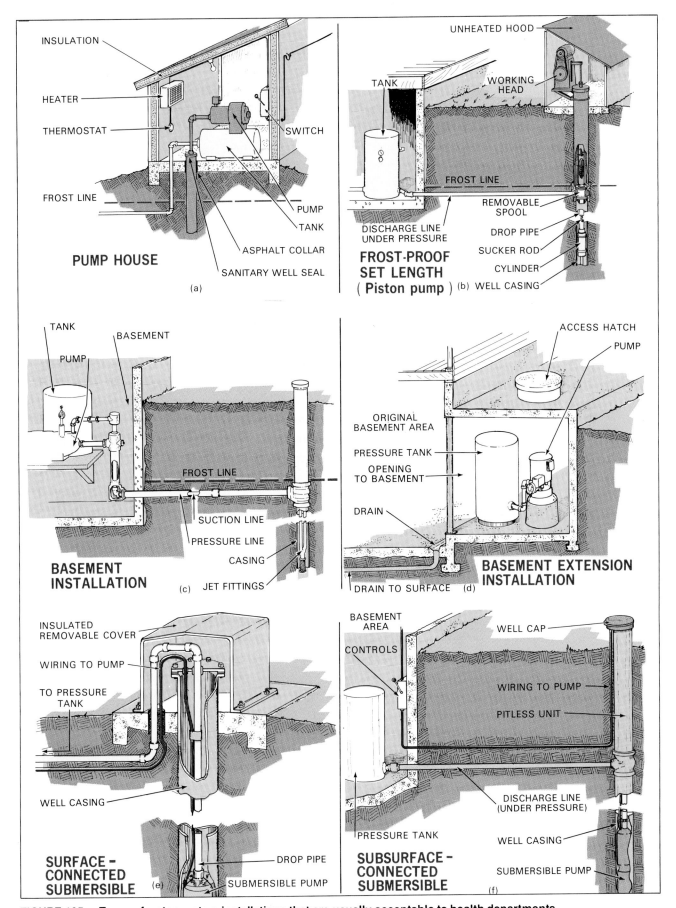

FIGURE 105. Types of water-system installations that are usually acceptable to health departments.

TABLE XV. ADEQUACY OF VARIOUS TYPES OF WATER-SYSTEM INSTALLATIONS

Factors to Consider	Pump House (Fig. 106a)	Frost-Proof Set Length (Piston pumps) (Fig. 106b)	Basement Installation (Fig. 106c)	Basement Extension (Fig. 106d)	Surface Connected Submersible (Fig. 106e)	Subsurface Connected Submersible (Fig. 106f)	Pump Pit (Fig. 107a)	Offset Pump Pit (Fig. 107b)	Pit-Type Frost-Proof Set Length (Fig. 107c)	Subsurface Sealed Casing (Fig. 107d)
Protection from Surface Water Entering Well	Excellent	Excellent	Excellent, if basement is well drained. Rating of "excellent" assumes the well casing will extend 8 to 24 inches above ground level and that the top will be sealed.	Excellent	Excellent	Excellent	**Usually poor.** Even if the well casing extends 8 to 24 inches above the floor, the pit can partially fill with water that may enter under the cover slab, back through the drain or enter through a crack in the wall or floor.			**Usually poor.** It is difficult to make well cap water tight.
Adequacy of Weather Protection	Excellent, if tightly constructed, well insulated and heated if necessary.	Excellent	Excellent	Excellent	Excellent	Excellent	Excellent	Excellent	Excellent	Excellent
Adequacy of Drainage	Excellent, if water can be drained to ground surface.		Excellent, if basement is well drained.		No problem	No problem	**Usually poor.** When water table rises, water may enter pit through the drain and cause flooding. Even when drain extends to a surface outlet, slime and dirt tend to clot it until it becomes ineffective.			Excellent for equipment if basement is drained. Poor for well.
Adequacy of Ventilation	Excellent, if properly constructed for ventilation.	No problem	Satisfactory	Satisfactory	No problem	No problem	Poor	Satisfactory	Satisfactory	Satisfactory
Ease of Cleaning Enclosure	Excellent	No Problem	Excellent	Satisfactory	No problem	No problem	Difficult	Poor	Poor for pit	Excellent
Ease of Servicing Equipment	Excellent, if house can be removed, or if complete roof or roof section can be removed for major servicing or equipment replacement	Excellent, for pump head. Good for water units.	Excellent, if equipment is not crowded.	Excellent, if equipment is not crowded.	Excellent	Satisfactory	Fair to satisfactory	Excellent	Satisfactory	Excellent
Ease of Servicing Well		Excellent	Satisfactory	Excellent, if overhead cover or complete roof is removable.	Excellent	Excellent	Satisfactory, if top is removable.	Satisfactory	Satisfactory	Poor

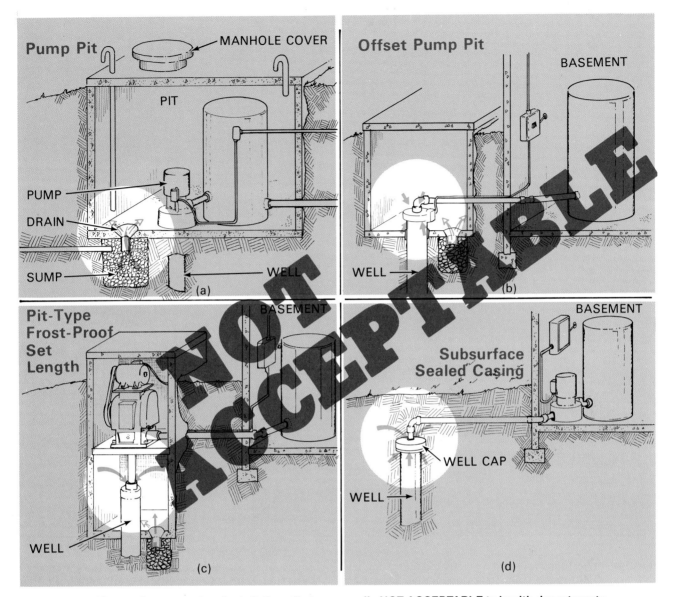

FIGURE 106. Types of water-system installations that are usually NOT ACCEPTABLE to health departments.

EASE OF SERVICING THE WELL

Provision for servicing the well is important if you are providing a pump house, or extending your basement to include the well area. After a period of use, your well may provide less and less water. In that case, it may be necessary to employ a well driller to deepen or redevelop your well. This means that the **top of the structure,** or at least part of it, **must be removed** in order to reach the well with the necessary tools to check and correct the well condition.

With large pump houses, provision is usually made for an *access hatch* in the roof (Figure 88 inset) or for *removing the complete roof*. Small pump houses are usually designed to *remove the complete house*.

If you have a basement extension, either the *roof can be removed* or an access hatch is provided in the roof directly over the well (Figure 105d).

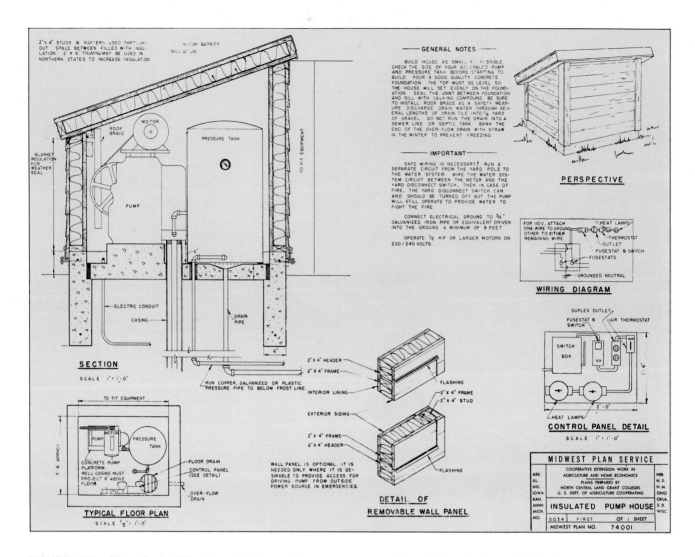

FIGURE 107. Plans for building insulated pump houses are available through your State Cooperative Extension Service. As an example, here is Midwest Plan No. 74001. It can be adapted to different sizes of installations. A somewhat similar plan is available from the University of Minnesota, Dept. of Agricultural Engineering, St. Paul, Minn. It is plan sheet M116. Complete specifications are provided with each.

VI. Planning the Piping Installation

Thus far, you have determined what use you will make of water, and you have selected the size pump and water storage which it takes to supply those uses. Your next job is to plan the location of outside water-use outlets—those inside of buildings are located as part of the building plan—and then to provide an adequate piping system to deliver the water to the buildings and to the yard hydrants.

Planning the location of water-use outlets is important for both convenience and fire control. But, providing a pipe distribution system that is adequate to meet your water needs is vital for a successful system. If your piping system is not adequate, much of your planning up to this point and your investment will be lost. The reason: the water supply you need and expect to get at the various outlets will be choked off because of a piping system that is too small.

If the piping system is inadequate to begin with, it will become more inadequate as you increase your uses of water, and as the supply pipes begin to corrode (roughen) on the inside, or collect deposits.

In order to plan an adequate piping system, you will need to learn:

A. Where to Locate Outside Water Outlets
B. How to Plan the Piping Layout
C. What Kind of Pipe to Select
D. How to Select the Proper Supply-Pipe Size
E. What Pipe Protection to Provide

These are the headings of the discussions that follow.

A. Where to Locate Outside Water Outlets

Standard recommendations have not been developed as a guide for locating outside water outlets around the home and on the farm. But they involve two important factors:
— Locating outlets for convenience
— Locating outlets for fire protection

LOCATING OUTLETS FOR CONVENIENCE

This discussion is limited to planning the water outlets outside of the home and outside of the service buildings on a farm.

Location of water **outlets in the home** is determined largely by the house plan and the location of the conveniences and equipment requiring water connections.

When planning outlets outside of your home it is good to have one outlet for connecting a hose located about every 100 feet apart as you measure **around the outside of the house**. As shown in Figure 108, this allows a

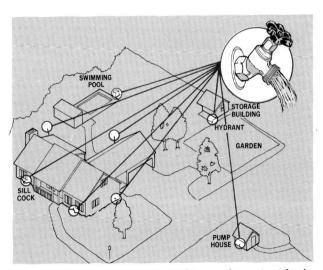

FIGURE 108. Outside water outlets are important for irrigation as well as for cleaning walks, porches, driveways and for other hose-cleaning jobs.

FIGURE 109. Some of the more common uses of water in farm service buildings.

50-foot hose to reach any place around the edge of the house for watering shrubs as well as for hosing down porches, steps and sidewalks. These are usually sill cocks as shown in Figure 108 inset.

The need for **other water-use outlets** will vary with the applications you have in mind and with how highly you value convenience in having water available. In Figure 107, it is evident that there should be water available for filling the *swimming pool,* for maintaining the water level, and possibly for showers in the dressing room.

A hydrant close to the storage building and garden would add to the convenience for *garden watering* and would also be convenient for *cleaning tools and equipment* that may be housed in the storage building. A hydrant is suggested to avoid a freezing problem. How a hydrant works is shown in Figure 128.

A *sill cock on the pump house* provides a convenient water outlet for that portion of the yard area.

For the area shown in the illustration there would probably be little need for more outlets. This approach to planning makes water available almost anyplace about the home with a 50-foot length of hose.

If you have a **farm** with one or more service buildings, you will need to consider them individually. If you have new buildings, the location of water outlets in and around them was probably considered when the building was designed. If you are planning water outlets in existing buildings, the brief explanation that follows may be of help.

Figure 109 shows a layout of service buildings surrounding a home and possible uses for water in each of them. There is not enough space in the illustration to show all possible uses so you may wish to check others listed in Table IX as a reminder.

Location of water outlets in the **barn** will be influenced by how the barn is used. If you have a *stanchion-type dairy barn* and keep your cattle in the barn all night, it is good to provide automatic drinking cups—one for each two cows. Then provide an outlet for connecting a hose that will serve about 50 feet in each direction. This will take care of flushing the floor and gutter.

If your barn is for *loose housing of cattle,* you will need an outlet for a watering tank or for one or more automatic waterers. Be sure to install additional water outlets in any area where you have calf stalls or maternity stalls.

If you have a *milking parlor,* you will need at least one outlet for a hose connection to clean the floor. A warm-water spray connection for washing udders saves time and conditions the cows for milking. An outlet for each two cows is desirable for this.

If you are using a *milking machine,* a rinse fountain supplied with hot and cold water enables you to keep the teat cups clean and aids in keeping down udder infection.

In the *milk room* you will need water outlets to supply the utensil washing vat, and a separate hose-connection outlet for cleaning the floor. If a toilet and lavatory are to be installed, provide outlets for them.

In a *hog house,* plan at least one outlet for supplying each automatic, permanently-connected waterer and one or more hose outlets for floor cleaning and supplying water for movable hog waterers.

Add enough hose-connection outlets so that all parts of the housing and *feeding floor* can be reached with 50 feet of hose for cleaning.

If you are planning to install a *hog wallow,* allow an additional outlet for it.

LOCATING OUTLETS FOR FIRE PROTECTION

Here are the recommendations of the USDA regarding planning for *fire protection*:[36]

> Hydrants and hose valves should be located in and about the buildings to be protected at 100 foot intervals. Normally 50 feet of hose would reach any point where a fire occurred. If a fire occurred near a hydrant, so that it could not be used, 100 feet of hose would be needed to reach that point.

The recommendations of the National Fire Protection Association are essentially the same.[31]

For more complete access to water in case of fire, one other point needs to be considered. When you are depending on water outlets in a building or on the outside of a building you may find **some outlets are not useable because of too much smoke and heat** when you are that close to the building. A better arrangement is to provide hydrants located about 50 feet away from the buildings and in a position where some of them can be used to fight fire on more than one building.

B. How to Plan the Piping Layout

In this discussion, planning of the piping layout has to do with the piping between the pump house and the residence and nearby structures. Or, if you have a farm, the piping between the pump and all of the buildings to which water is to be supplied.

Your job is to determine **how much water demand there will be at each building, or at each outside hydrant,** then determine the most satisfactory pipe layout. This kind of planning helps keep down piping and trenching costs, and keeps down water-pressure loss. The latter will also be considered when you select the kind and size of pipe to use.

The procedures for planning your piping layout between buildings are given under the following headings:

— Determining water demand at various locations
— Determining where to place the pipe line(s)

DETERMINING WATER DEMAND AT VARIOUS LOCATIONS

In the previous discussion you considered **what uses** would be made of water in the various buildings and at outside hydrants. Now you need to determine **how much water** you will need to supply to each building and to each independent hydrant to meet the peak demands of each. The procedures for determining the peak demands are similar to those you used in arriving at the peak demand for the pump. The difference is that the demand must be determined separately for each location.

In arriving at the peak demand for the pump it made no difference where the various uses were located. Your job was to make sure you had all of them listed so they could be considered in arriving at the peak demand on the pump. Now, for determining the water demand at each location, the *location of each use* is important. This is the information you need so you can (1) lay out an efficient piping system, and (2) later determine the pipe sizes that are needed to deliver enough water to meet the demands of each.

FIGURE 110. When figuring the water demand for the home, list all uses and the peak demand allowance for each.

To determine the **demand for each building and for each outside hydrant,** proceed as follows:

1. *List all of your home uses and the "peak demand allowance" for each.*

    Use Table IX, column 1.

    As an example, assume that the uses are the same as those listed when you were figuring the peak demand for the pump (Figure 110).

2. *Determine which use has the greatest fixture-flow rate.*

    Check column 2, Table IX, for the greatest *fixture-flow* rate for the water uses you have listed.

    Usually, a tub-and-shower combination or a clothes washer has the highest flow rate. However, if you have either a **neutralizing filter or an iron-and-sulfur filter,** the flow rate for one of these is probably higher. Neither of these is listed in Table IX. You will need to check the specification for either of these to determine the exact flow rate since it varies for different sizes.

    The **purpose of using the flow rate** for the heaviest water user is to make sure you provide plenty of water, so that when it is in use it will not "rob" the others that may be in use at the same time.

    In the example, the tub-and-shower combination and the clothes washer both have a flow rate of 8 gpm. Only one needs to be considered.

3. *Substitute the "fixture-flow-rate" figure of the heaviest water use for its "peak-demand-allowance" figure.*

    Figure 110 shows an example.

FIGURE 111. The fixture-flow-rate figure for the heaviest use replaces the peak-demand-allowance figure.

4. *Add any competing demand such as lawn or gardening sprinkling, if supplied from home plumbing system, then total (Figure 112).*

    In this example, assume that **one yard sprinkler** will be used and that the water will be supplied from the home plumbing. Since it may be used continuously for several hours, it is considered a competing demand and is listed at its fixture flow rate.

    The **swimming pool** has not been added to the demand. After filling it the first time, very little water is needed to maintain the water level.

5. *Determine the water demand for other buildings (Figure 113).*

    Use the same steps in arriving at the demand for each building that you used in determining the demand for the home. Use Table IX as reference.

    Note that the *first figure for each building* shows the "fixture-flow-rate" since that is the heaviest use. The others are "peak-demand-allowance" rates.

6. *Determine the demand for outdoor water uses such as yard hydrants and swimming pool.*

    Use Table IX as reference.

If you total the demands for the home with those for the service buildings and yard hydrants, you will find the total is much greater than the demand you figured

124

FIGURE 112. If lawn or garden sprinkling is supplied through the home plumbing system add the demand for it.

for the pump. The reason is that the **peak demand changes during the day** from one building, or location, to another as work progresses. For example, the capacity needed for hose-cleaning floors was included for the poultry house, for the barn, for the milking parlor and for the hog house. In actual use, floor flushing would probably not be done at more than one location at any one time.

DETERMINING WHERE TO PLACE THE PIPE LINE(S)

When planning a pipe-line layout, you will usually find there are several ways it can be done. Your first thought may be to develop the layout that takes the least amount of pipe length and the least amount of trenching. But you may also need to consider such matters as cutting through concrete or asphalt paving, and locating cutoff valves so as to cause the least disruption of water service when one of the pipe lines is being serviced.

The following procedure will help you keep these points in mind as you develop your layout:

1. *Make a sketch of your buildings and grounds (Figure 114).*

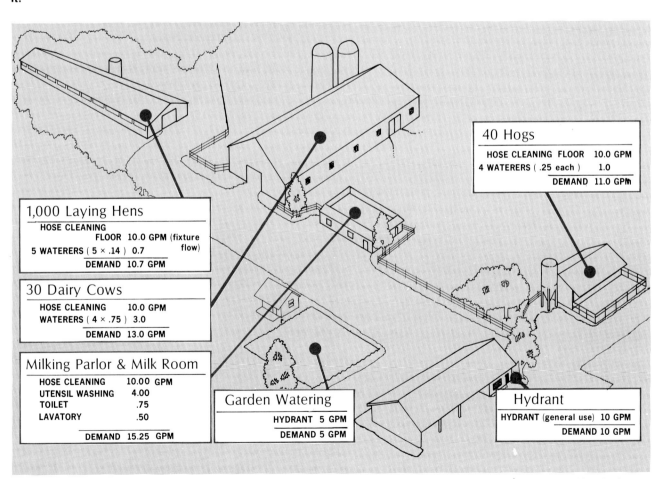

FIGURE 113. Figure the demand for the farm service buildings in the same way you arrived at the demand for the home.

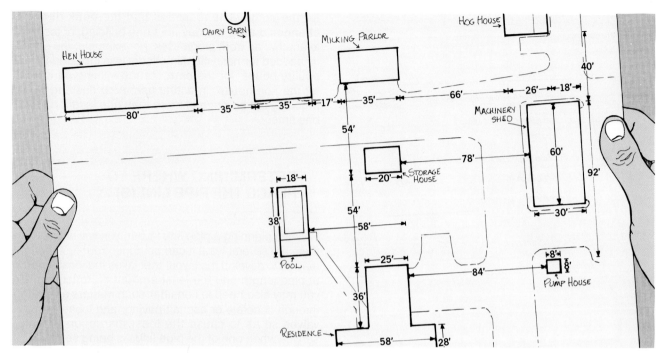

FIGURE 114. Make a sketch that shows your buildings and the critical distances involved for planning your piping layout.

Use a large sheet of paper and sketch in the buildings to show approximately where each building is located in relation to the others. It is helpful to make the sketch to scale. Use ⅛ inch or ¼ inch to equal one foot.

2. *Measure the distances between buildings as well as building widths and lengths, and record on sketch (Figure 114).*
3. *Show the water demand at each location in the sketch (Figure 115).*

Show the demand you figured for each building on the sketch. Also, show the demand you figured for each outdoor hydrant.

4. *Draw a double line from the pump house to the location of greatest demand (Figure 115).*

In the example, the home has the greatest demand. Note that the double line, representing the proposed pipeline, is labeled "A."

5. *Place a double line to the next closest major demand (Figure 115).*

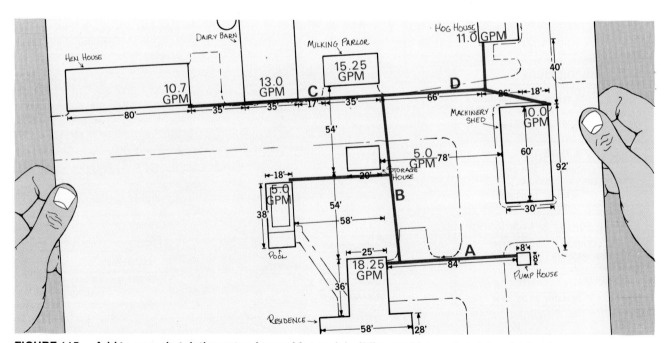

FIGURE 115. Add to your sketch the water demand for each building and for each outdoor hydrant.

126

Since the milking parlor and milk room location is the next closest site of a major demand, extend the double line to it. You may extend it from either proposed pipeline "A" near the house, or from the pump house. The distance is somewhat less from the "A" line to the milking parlor. Also, the garden hydrant can be supplied enroute without extra piping to serve it. Consequently, for now, proposed pipeline "B" is connected to pipeline "A."

6. *Draw double lines to the next closest location of a major demand (Figure 115).*

The next major demand location is the barn, then the poultry house as you move to the left of pipeline "B." This is labeled "C." In the opposite direction from pipeline "B" is the hog house, then the machinery storage and shop. Extend double lines from "B" to reach them.

If your plan has other major demand locations, continue the double line to them.

7. *Draw lines from line "A," "B," "C" and "D" lines to any other water-use locations (Figure 115).*

Only the swimming pool remains to be connected. This can be reached easily from proposed pipeline "B."

8. *Check layout for ease of installation and maintenance.*

Now that you have completed your preliminary piping layout, you may find that changes need to be made to meet existing conditions. For example, suppose the *drive to the house is paved,* including the turn around. To avoid cutting through so much paving, line "A" could be lowered on the sketch from its present position to one where it would miss most of the paved area (Figure 116).

At the same time, you might wish to consider running pipeline "B" from the pump house to the general vicinity of the garden hydrant and the pipe connection to the swimming pool. That would avoid cutting through the paved section next to the garage.

Another point to consider is the **location of cut-off valves.** With the first layout there would normally be a cut-off valve at the pump house where it serves pipeline "A." If for any reason the "B," "C" or "D" lines needed to be repaired, the one cut-off valve at the pump house would discontinue water service to the home and all the rest of the buildings. Consequently, it would be desirable to have a cut-off valve on pipeline "B" as shown in Figure 116.

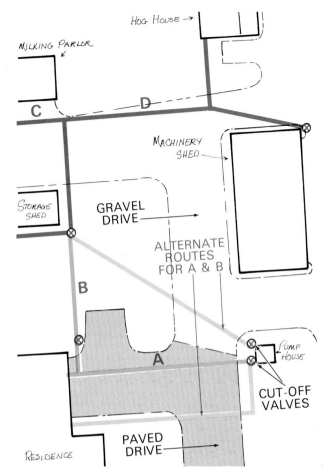

FIGURE 116. Review your plan to see if it needs to be changed to provide easier installation and provision for better water service.

If you moved pipeline "B" to the position shown in Figure 116, the cut-off valve for it could be located at the pump house. This would have the added advantage that turning the water off to the house would not affect the water supply to the service buildings.

The cut-off valves could easily pay for themselves if water to dairy cows or chickens would have to be turned off for as much as a day while repairs are made.

If you are operating a large farm, you may be justified in installing cut-off valves on other branch lines to your service buildings, or a cut-off valve at each service building. This is determined by how important a water service interruption would be in case it lasted for several hours.

C. What Kind of Pipe to Select

In reaching your decision as to what kind of pipe to use for your underground supply lines, you first need to understand about the different kinds of piping materials and then learn the different qualities of each. This information is included under the following headings:
— Kinds of piping
— Probable life expectancy
— Resistance to corrosion
— Resistance to deposits forming inside of the pipe
— Effect of freezing
— Safe working pressure
— Resistance to puncture and rodents
— Effect of sunlight
— Effect on water flavor
— Lengths available
— Comparative weight
— Ease of bending
— Ability to conduct electricity
— Comparative cost
— Ease of installation

The information for all of these factors is given in Table XVI except for the one dealing with "Ease of Installation." It is discussed separately.

If you are not acquainted with the kinds of piping materials commonly used for water distribution in individual water systems, the discussion that follows will help you understand the table.

KINDS OF PIPING

The kinds of piping most readily available for your use are the ones discussed in Table XVI. They are:
1. Galvanized-steel pipe (standard weight)
2. Copper pipe or tubing
3. Plastic pipe

Galvanized-steel pipe has a protective coating of galvanizing material (zinc) which greatly increases the life of the pipe as compared to black pipe. For that reason black pipe is not included in this discussion. Before plastic pipe was developed and improved, galvanized-steel pipe was used almost exclusively for underground supply lines in rural areas.

Galvanized pipe is rapidly being replaced by plastic pipe.

Copper pipe is available in types "K" and "L." "Types" refers to pipe weight. Type **K** is for heavy duty such as for pump suction lines or for underground use where there will be considerable outside pressure such as under driveways. Type L is standard weight copper pipe for general use both underground and in buildings. Of the two types, Type L is most widely used in rural areas.

Copper pipe is seldom used except indoors.

Both types are available in "hard-tempered" or "soft-tempered" form. *Hard tempered is rigid.* It comes in straight lengths of 12 feet and 20 feet. It is desirable for use inside buildings where it will be exposed. You can fit it closely to walls or ceilings. It needs very little mechanical support to keep it in position, compared to flexible tubing. *Soft-tempered tubing* is excellent for underground use and for fishing between existing walls in old buildings.

Plastic pipe is a comparative newcomer. However, it is finding wide use in rural areas for underground supply lines. Plastic pipe is made from a variety of different plastic materials. The kind most available to rural areas is made either of polyethylene (PE) or of polyvinyl chloride (PVC). The initials are commonly used to identify each of them.

PE plastic pipe is usually black. It comes in large coils and is reasonably flexible under warm conditions.

PVC piping material is usually white or gray in appearance. The pipe comes in 10-foot lengths and is quite rigid. It is more resistant to crushing or puncturing than PE piping, and it will stand somewhat higher water temperatures. However, water temperatures are usually no problem with underground piping.

TABLE XVI. RELATIVE MERITS OF DIFFERENT KINDS OF PIPING MATERIALS FOR SUPPLY LINES

Factors to Consider	Galvanized Steel (3 oz. coating min.)	Copper Type K (heavy duty)	Copper Type L (standard)	Plastic
Underground soil corrosion—Probable life expectancy (1)	30 plus yrs. under most soil conditions. (If no corrosion inside pipe, life could extend to 100 yrs. or more.)	40-100 yrs. under most conditions.	30-80 yrs. under most conditions.	Experience indicates durability is satisfactory under most soil conditions.
	Waterlogged soils under most conditions—12-16 years. May be less than 10 yrs. in very high acid soils.	14-20 yrs. in high sulfide conditions. May be less than 10 yrs. in cinders.	12-14 yrs. in high sulfide conditions. May be less than 10 yrs. in cinders.	
Resistance to corrosion inside pipe.	Will corrode in acid, alkaline and hard waters or with electrolytic action. (2)	Normally very resistant. May penetrate rapidly in water containing free carbon dioxide.		Very resistant
Resistance to deposits forming inside pipe.	Will accumulate lime deposits from hard water. (2)	Subject to lime scale and encrustation from suspended materials.		Resistant, but occasional deposits will form. (3)
Effect of freezing	Bursts if frozen solidly.	Will stand mild freezes.		PE—will stand some freezing. PVC—will stand mild freezes.
Safe working pressures (lbs. per sq. in.)	Adequate for pressures developed by small water systems.	Adequate for pressures developed by small water systems.		Working pressures at 73°F. PE 80 to 160 PVC 180 to 600
Resistance to puncturing and rodents.	Highly resistant to both.	Resistant to both.		PE—Very limited resistance to puncture and rodents. PVC—resistant
Effect of sun-light	No effect	No effect		PE—Weakens with prolonged exposure— PVC—Highly Resistant
Effect on water flavor	Little effect	Very acid water dissolves enough copper to cause off flavor.		Little effect (4)
Lengths available	21 ft. lengths	Soft temper: 60-ft.—100-ft. coils up to 1″ diameter. 60-ft. coils above 1″ diameter. Hard temper: 12- and 20-ft. lengths		PE usually in 100-ft. coils, or longer PVC usually in 20-ft. lengths
Comparative Weight (Approx. lbs. per foot) Inside diameter				
½	.85 lb.	.34 lb.	.29 lb.	.06 lb.
¾	1.13	.64	.46	.09
1	1.60	.84	.66	.14
1¼	2.27	1.04	.88	.24
1½	2.72	1.36	1.14	.30
2	3.65	2.06	1.75	
2½	5.79	2.93	2.48	
Ease of bending	Difficult to bend except for slight bends over long lengths.	Soft temper bends readily, will collapse on short bends. Hard temper difficult to bend except for slight bends over long lengths.		PE Bends readily, will collapse on short bends. PVC Rigid Bends on long radius
Conductor of electricity	Yes	Yes		No
Comparative cost index Pipe size (in.)				PE PVC
½	10	21	17	3 3
¾	13	39	28	5 4
1	18	54	42	9 5
1¼	24	66	59	14 7
1½	28	87	74	18 10
2	39	130	110	27 15

(1) Derived from studies reported by Dennison, Irving A. and Romanoff, Melvin, "Soil-corrosion Studies, 1946 and 1948: Copper Alloys, Lead and Zinc," Research paper RP2077, Vol. 44, March 1950 and "Corrosion of Galvanized Steel in Soils," Research paper 2366, Vol. 49, No. 5, 1952, National Bureau of Standards, U.S. Dept. of Commerce.

(2) It is possible to greatly reduce corrosion and prevent lime scale in steel pipe by adding a phosphate material. It coats the inside of pipes, as well as the lining of all connected equipment. Prevents further lime scale and greatly reduces corrosion.

(3) Jones, Elmer E. Jr., "New Concepts in Farmstead Water System Design," Am. Society of Agricultural Eng., paper No. 67-216, 1967.

(4) Tiedeman, Walter D., "Studies on Plastic Pipe for Potable Water Supplies," Journal American Water Works Association, Vol. 46, No. 8, Aug. 1954.

(5) U.S. Environmental Protection Agency, 1982. In "alkaline" waters, studies have indicated equal or superior performance of galvanized pipe to that reported for copper. In fact, in waters with low to moderate pH and high alkalinity, galvanized pipe would often dissolve less than copper pipe. Copper plumbing only **appears** to be more acid resistant than galvanized because of red water problems when the galvanizing has dissolved away. However, copper is quite soluble in any water that is even slightly acidic. Copper is probably more soluble in "hard" waters than galvanized, to the extent that the dissolved carbonates (HCO_3^-, CO_3^{2-}) attack copper to some degree, but tend to passivate zinc until very high carbonate levels are reached.

The type of phosphate material is very important. Many polyphosphate compounds will not only reduce scaling, but will also cause increased uniform dissolution of the metal pipe. The individual probably would not be aware of the difficulty without frequent metal analyses of his water.

High levels of chlorine are also detrimental to copper pipe.

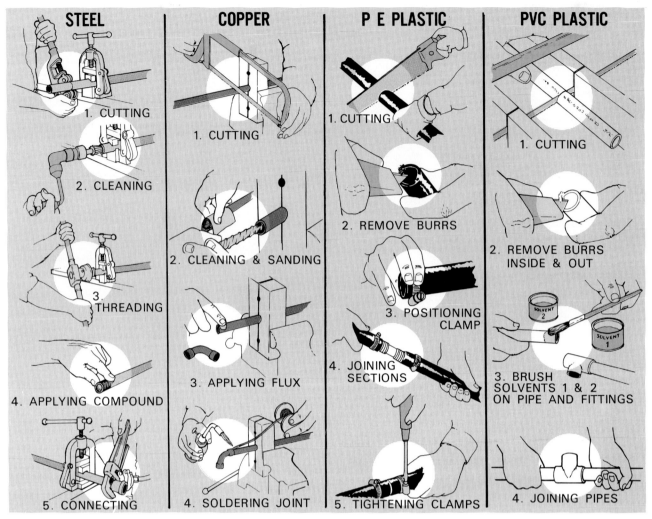

FIGURE 117. Steps in preparing and connecting steel, copper, and plastic pipe.

Both PE and PVC plastic pipe are available at different **pressure ratings.** The ratings most commonly used for PE pipe are 100 and 120 pounds per square inch (psi), and for PVC pipe, 200 psi.

Note in Figure 117 that PE and PVC pipe sections are fastened together differently.

Both kinds of pipe are available in *sizes up to 6 inches* in diameter or larger.

When you buy plastic pipe, you will notice several **markings** on it. These include pipe size, letters indicating the materials from which it was made—PE or PVC, working pressure, and possibly the letters "nSF." The nSF marking indicates the pipe has met the specifications of the National Sanitation Foundation Testing Laboratories, Inc. and is suitable for delivering drinking water.

Polybutylene pipe is a new type of plastic pipe that offers a number of advantages. It is bendable, easy to work with and install. Gasoline will penetrate it, however, so do not put polybutylene lines close to fuel storage tanks or under parking areas for cars, trucks, or farm machinery such as tractors or self-propelled implements, if at all possible.

CPVC and PB plastic pipe can be used for either hot or cold water. Follow manufacturer's recommendations.

EASE OF INSTALLATION

There are no time studies on how long it takes to install the different kinds of pipe. But Figure 117 shows the different steps in joining pipe sections of all three kinds. You can get some idea of ease of installation from it.

If you already have pipe of one kind installed and wish to change part of it to another kind, there are fittings available so you can make the proper connections from one kind of pipe to another.

D. How to Select Proper Supply-Pipe Size

Now that you have determined the kind of pipe you plan to use, your next decision is to arrive at what size piping to select to supply water to the various locations. There is sometimes a tendency to guess at the size. If the guessed size is too small—as it usually is—you will partially offset the planning you did in selecting the proper size pump. The size of the delivery pipes will be too small to deliver the amount of water you found you need at the various locations.

How a **small increase in pipe size** greatly increases water flow is explained in a Maryland bulletin[37] with the following example:

> A small increase in the size of pipe installed will cost a little more initially, but will mean much in increasing system efficiency and in providing for future needs.

RELATIONSHIP OF PIPE SIZE TO WATER FLOW

Pipe Size (inches)	Water Flow (gal. per min.)	Comparative Increase (per cent)
¾	4.0	100% (base)
1	7.5	187% (almost double)
1¼	16.0	400%

The above comparison was made at 2 lb. per square foot friction loss per 100 feet of pipe.

The amount of water a pipe will deliver at any one pressure is determined (1) by the size of the pipe as just shown, (2) by the smoothness of the inside surface, and (3) by the pipe length.

As shown in Figure 118a, water flow is affected by the **smoothness of the inside surface.** Even in a very smooth pipe, the water is held back somewhat as the water moves along the inside pipe surfaces. This is called "friction". If the pipe is rough on the inside surface (Figure 118b), the water flow is held back much more because of increased friction. This means that a smooth pipe, compared to a rough pipe of the same size, will deliver more water under the same pressure.

Although this discussion deals only with supply-pipe sizing, it is important for you to be aware that **pipe connections**—elbows, reducers, tees, unions, etc.—as well as valves, can add greatly to the friction problem.

The amount of friction caused by each of these units is expressed in how much pipe length it would take to provide the same amount of friction. The importance of this situation as it applies to individual water systems is brought out in the following observation:[2]

> Within 5 feet of the hydropneumatic tank there will normally be found from three to six elbows or tees and two valves. For 1-inch line with four elbows, two tees side outlet, and two gate valves, the equivalent length (in pipe) is 36.2 feet. With two full size globe valves it is 96.4 feet. With two undersize globe valves it may exceed 658 feet. . . .

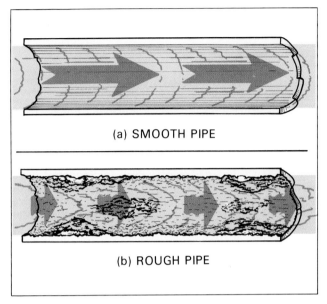

FIGURE 118. Effect of interior smoothness of pipe surface on flow of water. (a) A pipe with a smooth inner surface provides little resistance to water flow as it contacts the inside walls of the pipe. (b) A rough inner surface delays water movement, thus increasing resistance to flow.

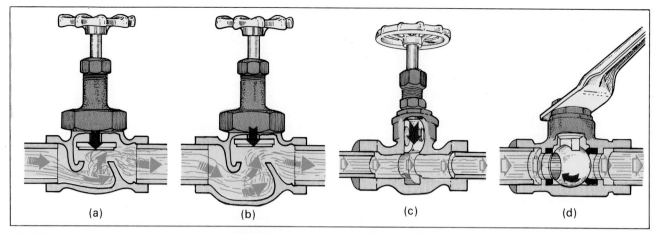

FIGURE 119. Types of cutoff valves. (a) Economy globe valve causes high resistance to water flow because of narrow passageways and sharp turns. (b) Globe valve with larger passageways has lower resistance to water flow. (c) Gate valve, when completely open, provides straight-line water-flow passage with very low resistance. (d) Ball valve. It also provides easy water passage and low resistance.

Figure 119 shows the common **types of cutoff valves** and how they work. Note in Figure 119a, that the valve opening is much smaller than the pipes connecting to the valve on either side. Also, the water must make 2 sharp turns in passing through the valve. These are the conditions that cause high friction. This type of valve may be equal to as much as 200 to 300 feet or more of pipe length. The same type of valve with larger passageways (Figure 119b) greatly reduces the friction but, even with larger passageways, the loss may be equal to 50 to 100 feet of pipe length. Either a gate valve (Figure 119c) or a ball valve (Figure 119d) greatly reduces friction because both of them provide a straight-line passage of water through the valve opening. If the valve opening is the same size as the pipe connected to it, friction loss may be as low as 2 to 10 feet of equivalent pipe length.

In the previous discussion you studied the various characteristics of steel, copper and plastic piping materials. One important factor was not discussed and that is *how much friction water will encounter* as it flows through piping made of each of these materials. This has been summarized into charts which you will use as you determine the pipe sizes needed to provide for underground water service to your home and to other buildings.

The procedures you use for pipe sizing are affected by whether you have *only one building or hydrant* to serve with a pipe line, or *have two or more to serve*. Where there are two or more, you will take into account the non-competing uses at the different locations much the same as you did earlier when you were planning the demand allowance for the pump. This time the demand allowance reflects on the size piping you will need to install.

If you are interested in the piping being adequate to provide some **fire protection,** this will need to be taken into account with all of the supply lines.

The information you need and the procedures to follow in selecting the proper supply-pipe sizes are included under the following headings:

— Determining pipe size to serve one location
— Determining pipe size(s) to serve two or more locations
— Determining pipe sizes for fire protection

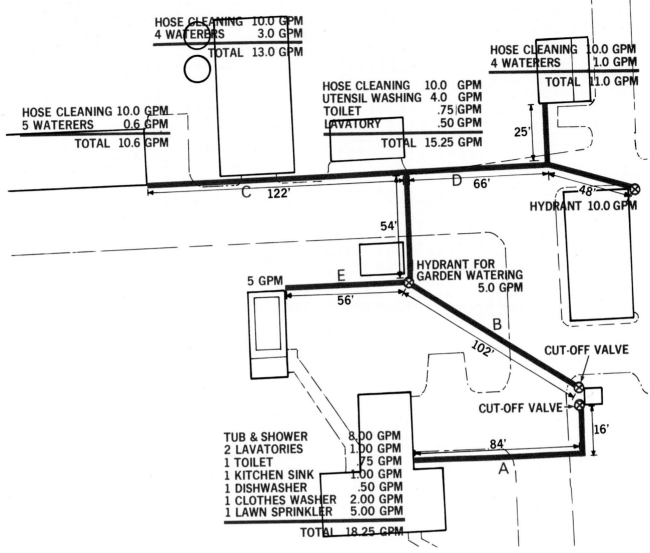

FIGURE 120. For pipe sizing, information is needed on pipe-line lengths and the water demand allowance for each use at each location.

DETERMINING PIPE SIZE TO SERVE ONE LOCATION

If the pipe line is to serve only one location such as between the pump and home, or between the pump and a hydrant or service building, proceed as follows:

1. *Determine the total demand allowance for that location.*

   Example: In Figure 120, one pipe line will supply the home only. The demand allowance for the home was determined earlier to be 18.25 gpm (Figure 112).

2. *Determine the distance from the pump house to the location.*

   Example: In Figure 120, the distance from the pump house to the home is 100 feet (84 ft. plus 16 ft.).

3. *Determine the proper pipe size to use.*

   Use the pipe size selection chart that applies to the kind of pipe you decided to use—Figure 121 for **plastic pipe,** Figure 122 for **copper pipe** or Figure 123 for **steel pipe.**

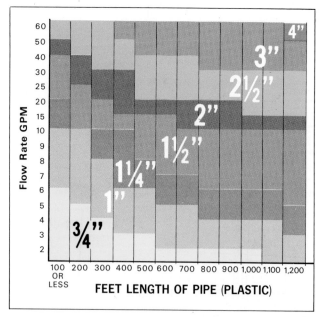

FIGURE 121. Pipe size selection chart for PLASTIC PIPE.*

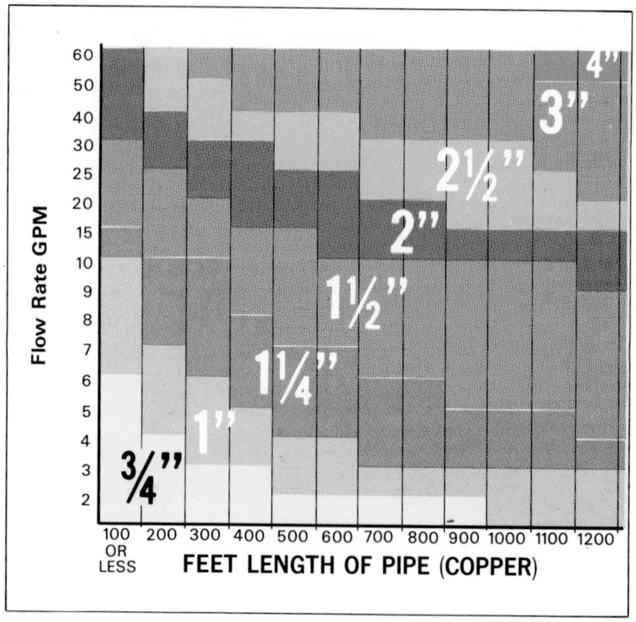

FIGURE 122. Pipe size selection chart for COPPER PIPE.*

Example: Assume you are using plastic pipe. The first flow rate greater than 18.25 gpm is 20 gpm (Figure 120). For a distance of 100 feet, the chart shows the pipe size should be 1¼ inches.

DETERMINING PIPE SIZE(S) TO SERVE TWO OR MORE LOCATIONS

Where two or more water-use locations are involved, the situation may vary from one where all uses are on one continuous line to one with branch lines, or even branch lines off of branch lines. The procedures that follow will enable you to deal with any of these situations:

1. *Indicate the water demand allowance for each use at each location on your piping layout chart.*

   It is helpful to lay out the information on your chart in a manner similar to that in Figure 120.

2. *Select the water use location farthest from the pump to start.*

   Example: In Figure 120 it is branch line "C" that extends to the poultry house.

3. *Compare the demand allowance at the most distant location with the demand allowance at any other location(s) served by the same line or branch line.*

   Example: Note in Figure 120 the cattle barn is also on branch line "C" that serves the poultry house. Note that the **demand allowance for each building is similar,** and that the buildings are located close to each other. With these conditions, use the full length of line "C" and figure the pipe size the same as if both buildings were on the end of the pipe line.

   If the **demand allowance at the cattle barn had been several times greater** than the demand allowance at the poultry house, pipe line length would then be figured from the cattle barn and only the poultry house demand allowance considered. This would save some cost by not having to extend to the poultry house the extra-large piping needed for the cattle barn.

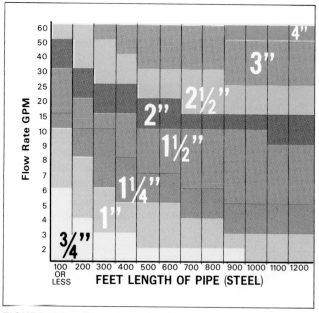

FIGURE 123. Pipe size selection chart for STEEL PIPE.*

---

*Developed from the 1971 National Standard Plumbing Code. It is based on fairly rough to rough surface conditions, a maximum flow rate of 4 feet per second and a maximum pressure drop of 2 psi for the lengths indicated.

4. *Take out any non-competing demands.*

   If this is a one-man operation, only **one hose-cleaning job** could be done at any one time. When the poultry house is being cleaned, there would be no similar demand at the cattle barn.

   The **waterers** at both locations are automatic and will compete with each other.

5. *Determine the water demand on that portion of the line.*

   Example:

Poultry house:	Hose cleaning	10.0 gpm
	5 waterers	0.6
Cattle barn:	Hose cleaning	(non competing)
	4 waterers	3.0
	Total	13.6 gpm

6. *Determine the pipe line length necessary to serve these locations.*

   Example: Note in Figure 120 that "C" is 122 feet long.

7. *Determine the pipe size needed.*

   Use the pipe-size selection chart for the piping material you are planning to use. If **plastic pipe,** see Figure 121. If **copper pipe,** see Figure 122. If **steel pipe,** see Figure 123.

   Example: Assuming plastic pipe is used, the table shows the next flow rate above 13.6 gpm is 15 gpm. The next figure higher than 122 feet is 200 feet. In the chart where the two lines meet from each of these figures, *1½-inch pipe is indicated.*

8. *Determine the pipe size(s) needed for other branch lines.*

   Use the same procedures as you just completed—steps 1 through 7.

   Example: Water demand **branch line "D" to sub-branches:**

Hog house:	Hose cleaning	10.0 gpm
	Waterers	1.0 gpm
Machinery shed and shop		(non-competing)
	Total	11.0 gpm

   Length of branch "D" is 92 feet.

   For plastic-type pipe, Figure 120 indicates a 1¼-inch pipe.

   Sub-branch to hog house:

Hose cleaning	10.0 gpm
Waterers	1.0
Total	11.0 gpm

   Distance from "D" to hog house—25 feet.

   For plastic pipe, Figure 120 indicates a 1¼-inch pipe.

   Sub-branch to machinery shed and shop:

Hydrant	10.0 gpm
Distance	48 feet
Pipe size	1 inch

   Branch from line "B" to the swimming pool:

Hydrant	5.0 gpm
Distance	56 feet
Pipe size	¾ inch

9. *Determine pipe size needed for main feeder line.*

   The non-competing demands at all locations enter strongly into sizing this pipe. For a one-man operation, only one hose-cleaning operation is included. For a two-man operation, two hose-cleaning demands would have to be included.

   Example: **Water demand allowance for main line "B"** to serve the various branch lines and milking parlor — assuming a one-man operation:

1 man, 1 hose-cleaning demand allowance	10.0 gpm
Other demands:	
Milking parlor:	
Utensil washing	4.0
Toilet	.75
Lavatory	.5
Poultry waterers	0.6
Cattle barn waterers	3.0
Hog waterers	1.0
Hydrant for garden watering	5.0
Swimming pool, infrequent demand	(non-competing)
Total demand allowance	24.85
Distance, 156 feet	
Pipe size, 1½ inch	

DETERMINING PIPE SIZES FOR FIRE PROTECTION

If you have followed the recommended procedures thus far in selecting your pump and sizing your pipe, you probably have the capacity to provide fire protection for your buildings. The USDA[36] recommends a **minimum pumping capacity of 10 gallons per minute, a pipe size (if plastic) of 1¼ inch or larger** for all underground lines and a complete piping system sized so there will be **no more than 2 pounds (psi) loss in pressure** between the pump and the hydrants or other water-use outlets.

In the example that has been used in this discussion, the pump far exceeds the 10 gallons per minute capacity, the underground supply piping system is 1¼- and 1½-inch pipe, and the tables you used for figuring pipe size are based on a pressure loss of 2 pounds (psi) or less. The only changes needed to meet minimum requirements would be to increase the pipe size of the smaller branch lines to 1¼ inches in diameter.

However, in the example it was determined that the pump should have a capacity of 26.35 gallons per minute (page 78). If you take full advantage of this capacity for fire fighting, you need to have the full 26.35 gallons per minute available at any point on your piping system. You then have full fire-fighting capability at each building.

In providing for maximum pump capacity to be delivered through any part of your underground piping system, use the pump capacity figure you determined in place of the demand allowance figure in arriving at pipe-line sizes.

If you provide larger pipe sizes for fire protection, you are also **providing for increased water use** in case your future operations expand. This could be a very worthwhile advantage.

Over a period of years the installation of larger pipe sizes may *save the cost of having to retrench* to remove the old pipe and the *additional cost of new pipe* to replace it.

E. What Pipe Protection to Provide

With your piping installation planned to take care of your home needs, or home and farm needs, your next decision is how to protect it. If your piping is not installed deeply enough, the **water may freeze in the piping system** during cold weather. Then, you are without water service until the weather warms enough to thaw the ground around the pipes. Freezing may also cause some of the piping to burst. Finding the damaged pipe and replacing it can be time consuming and expensive.

There are also problems of **mechanical damage** to pipe and hydrants caused by tractors, automobiles and large animals. How to handle these problems is discussed under the following headings:
— Protection from freezing
— Protection from mechanical damage

PROTECTION FROM FREEZING

If protection of pipes from freezing is a problem in your area, it will be well recognized by your installer. Figure 124 shows the **commonly accepted depths** for placing pipes underground to prevent freezing in different regions. These depth recommendations are only approximately correct. Soil conditions and high elevations affect their accuracy.

Since frost may not reach the depths shown in the chart every year, some **installers are inclined to use more shallow depths.** You may get by without much trouble if there is considerable water flow through the pipes daily, because pipes that are in use help protect themselves. The surrounding ground takes up heat from the water that is passing through the pipes and thus delays freezing. But severe cold, over a period of several days or weeks, will usually overcome this advantage.

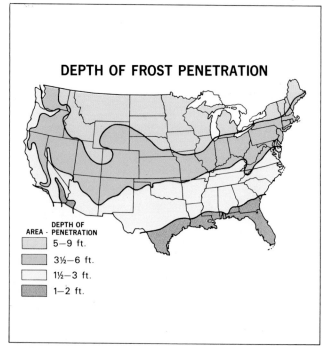

FIGURE 124. Greatest depths of frost penetration for various parts of the country. Piping in trenches should be placed below these depths.

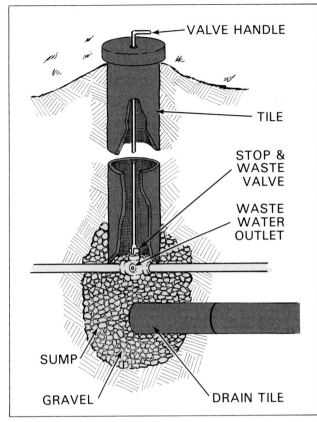

FIGURE 125. A stop-and-waste cock is used to drain unused pipe lines in winter to avoid freezing.

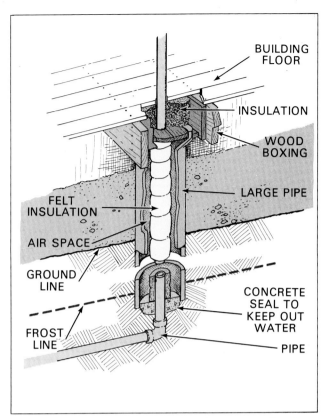

FIGURE 126. One method of protecting a supply line from freezing where it must extend through an open space between the ground line and a floor.

If you have a **pipe line that is not in use** during winter months, it is wise to drain it. A pipe line that is not in use does not supply heat to the surrounding soil and will freeze more readily than if it were in use.

You can take care of this problem by using a **stop-and-waste valve** for your cut-off valve in the supply line (Figure 125). With this arrangement, when the water is turned off, the valve automatically opens the waste port. Water from the nonpressure side of the line drains into the rock-filled sump around the waste outlet. Faucets fed by the line have to be opened so air may enter and make certain of complete drainage. Of course, your pipe line must slope from the faucets back to the stop-and-waste cock to be sure of complete drainage.

If your supply line is exposed to freezing where it passes from the **ground through an open space to the floor** of a heated building, you can provide protection by installing it as shown in Figure 126, or you can use a thermostat and electric heating element similar to that shown in Figure 127.

Supply lines that provide water for automatic poultry waterers or livestock waterers are commonly protected with an **electrical heating element and a thermostat** (Figure 127). A thermostat turns the electric current on and off as needed to supply protection against freezing.

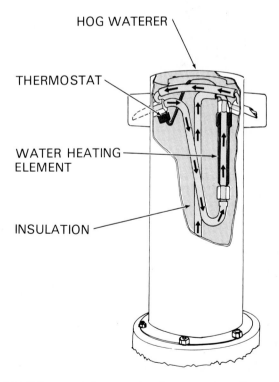

FIGURE 127. The supply pipe to automatic waterers is usually protected with an electrical heating element which is controlled by a thermostat.

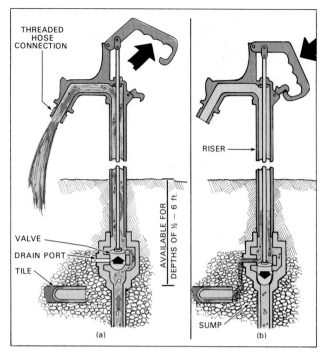

FIGURE 128. Frost-proof hydrants are designed for use where there is a freezing problem. (a) When the water is turned on, the drain port is automatically closed. (b) When the water is turned off, the drain port is opened. Water in the riser drains into the sump.

For hose connections in unheated buildings and for outside use, it is best to use a frost-proof hydrant (Figure 128). Each time you turn the water off, with a hydrant of this type, it automatically drains itself into a sump. Frostproof hydrants are available in different lengths so you can select one that will reach to any depth necessary to connect to your supply line.

Frost-proof hydrants have one bad feature — they can *cause pollution* if the sump is poorly drained. Water from outside the hydrant can flow back into your supply line while the drain port is open if the ground water around the sump rises above the level of the port hole. This has caused some health authorities to object to their use. If there is any probability of this happening, install a drain line that will allow the water to drain to an open outlet at a lower level.

Some effort is being made to design a hydrant that will overcome this pollution problem. At least one manufacturer is building a hydrant that provides for the water to drain from the riser into a water-tight container located below frost level. When the hydrant is used the next time, the water is removed from the container and exhausted along with the water coming from the supply line. This provides room for the next charge of water to drain from the riser when the hydrant is turned off. With this arrangement, no ground water can enter the hydrant.

For more information on hydrants, their use and abuse, see USDA Bulletin 2202, *Simple Plumbing Repairs for the Home and Farmstead,* December, 1972, available from your local Cooperative Extension Service office.

PROTECTION FROM MECHANICAL DAMAGE

Planning your outlets so they have mechanical protection is important for any location where **automobiles, trucks or tractors may run into them. Livestock also has a tendency to crowd them** out of position or break or nudge a valve open. For this reason, avoid putting a water outlet where livestock can have direct contact with it unless the outlet is to supply water to a tank or watering device which will protect it, or to which it can be securely fastened.

Where possible, keep your water outlets on the opposite side of the fence from livestock. If this is impossible, mount them along a post or other permanent protection that is easily seen and solidly placed. Figure 129 shows suggested ways of providing protection for water outlets exposed to mechanical damage.

If you are using **plastic pipe in shallow trenches,** it is good practice to provide mechanical protection where it extends under driveways. Heavy vehicles passing over the pipe can force a sharp stone or metal object through it if it is not protected. Place your plastic pipe inside a steel pipe to provide the necessary protection.

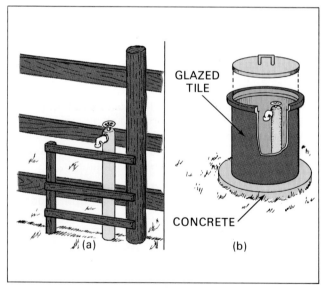

FIGURE 129. Ways of protecting water outlets from mechanical damage. (a) A fence post and board fence may supply adequate protection. (b) Protection for a completely exposed hydrant.

VII. Planning the Power Source

Thus far, your planning has involved getting enough water of the right quality delivered in enough quantity to your various water-use outlets to adequately meet your needs. If you have done a good job of planning thus far, your only remaining concern is to make certain your pump motor gets the energy and protection it needs. This involves three decisions:

A. What Power Source to Use.
B. What Wiring to Use.
C. What Electrical Protection to Provide.

The information for reaching these three decisions is included in the following discussion.

A. What Power Source to Use

Electricity is by far the most popular power source. But with the advent of high fuel costs, many people are looking back at the use of windmills, hydraulic rams and gravity systems. From this section, you will be able to **select the power source** best suited to your needs.

Power sources are discussed under the following headings:
1. Electrical Power for Pumping Water.
2. Windmills for Pumping Water.
3. Hydraulic Rams for Pumping Water.
4. Other Power Sources.

1. ELECTRICAL POWER FOR PUMPING WATER

Electricity is by far the primary source of power for pumping water. It is convenient, safe and economical. Electricity is also available to most places.

Electric motors for pumping water come in different types and sizes depending on your pump design (Figure 130). Generally, they are fractional horsepower capacitor type motors.

They may be either horizontal or vertical, depending on your installation needs.

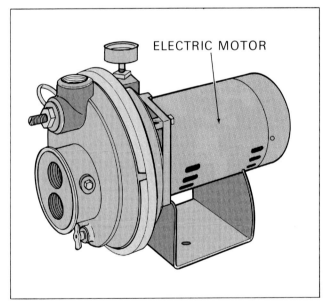

FIGURE 130. **A water pump using an electric motor for power.**

2. WINDMILLS FOR PUMPING WATER

The use of windmills for individual water systems is becoming more prominent now with the cost of electric power increasing. New designs are now available that will operate on low wind velocities.

With windmills, it is necessary for you to have a storage tank or reservoir for times when wind is not sufficient (Figure 131).

FIGURE 131. A windmill and water storage tank installation.

FIGURE 132. A hydraulic ram may be used to lift water.

3. HYDRAULIC RAMS FOR PUMPING WATER

The hydraulic ram is a cheap source of water power if you have the proper conditions (Figure 132). Water from a stream or spring may be pumped by a ram from a lower elevation to a higher elevation. Water flows down the drive pipe and develops power due to its weight and velocity. This power then is converted by design of the ram to power for lifting a part of the water to a higher elevation.

4. OTHER POWER SOURCES

Other power sources for pumping water are used depending on the need and availability. They include internal combustion engines and solar powered generators.

B. What Wiring to Use

If you use a feeder circuit with wires that are too small to adequately supply the electrical needs of the pump motor, you lose in two ways: (a) the useful life of the motor is shortened because of overheating, and (b) much of the electrical energy is lost in the form of heat in the feeder circuit (Figure 133a).

Wire size is determined not only by the energy needed by the motor but by the distance the energy is conducted through the wires and the voltage you plan to use.

The **type of conductor** you use is determined by whether or not the wires are to run overhead between buildings, or underground, or are to be extended through a building.

In making the proper wiring selection, you will need to consider the following factors:
— Wire size needed
— Type of conductor needed

WIRE SIZE NEEDED

Where individual water systems are used, the power supplier usually has **two voltages available**—120 volts and 240 volts. Unless you have a very low capacity pump you are almost certain to need 240 volts to operate your pump motor satisfactorily. Small motors of ⅓ horsepower or less can be connected to 120-volt circuits, but you will usually get better motor operation if it is connected to a 240-volt circuit. Motors of that size can often be operated on either voltage depending on how the connections are made on the motor.

There are three **reasons why a 240-volt circuit provides better service:** (a) the current flow (amperes) is only ½ as great with 240 volts as with 120 volts, which keeps down heat loss in the circuit and assures better motor operation, (b) you can usually use a smaller wire size, and (c) when your motor starts, there is little or no dimming effect on your lights.

After selecting the voltage you plan to use, you can then decide between **copper and aluminum conductors.** Copper has less electrical resistance to current flow than aluminum, which often means you can use a size smaller conductor for your circuit. But aluminum conductors are lighter weight and often cost less even if you use a size larger.

If your pump is one horsepower or more, use three-phase power. It is more efficient than single phase.

With these points in mind, check the motor horsepower on your pump and the distance from your electric meter to your pump motor. *For copper, use Table XVII to determine the wire size you need. For aluminum, use Table XVIII to determine the wire size you need.*

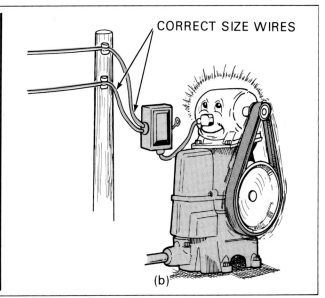

FIGURE 133. (a) A pump motor with feeder wires that are too small is "underfed" with electrical energy. Much of the energy is lost in the form of heat in feeder wires. (b) Wires of the right size help assure good pumping service and long motor life.

TABLE XVII. COPPER CIRCUIT-WIRE SIZES FOR INDIVIDUAL SINGLE-PHASE MOTORS
(Based on three per cent voltage drop on full-load current)

Motor size (h.p.)	Distance from Meter to Motor (feet)								
	50	75	100	200	300	400	500	600	700
120 Volts	Wire gage number								
¼	12	12	12	12	10	8	6	6	6
⅓	12	12	12	10	8	6	6	6	4
240 Volts									
¼	12	12	12	12	10	8	8	6	6
⅓	12	12	12	12	10	8	8	6	6
½	12	12	12	12	10	8	8	6	6
¾	12	12	12	10	8	6	6	6	4
1	12	12	12	8	6	6	4	4	3
1½	12	12	12	8	6	6	4	4	3
2	12	10	10	6	4	4	3	2	2
3	12	10	8	6	4	3	2	1	0
5	10	8	6	4	2	1	0	00	00

(If overhead wiring is used and span is more than 50 feet—use 8-gage wire for strength; or larger if needed for motor. Spans under 50 feet can use 10-gage wire unless larger size is needed for motor.)

TABLE XVIII. ALUMINUM CIRCUIT-WIRE SIZES FOR INDIVIDUAL SINGLE-PHASE MOTORS
(Based on three per cent voltage drop on full-load current)

Motor Size (h.p.)	Distance from Meter to Motor (feet)								
	50	75	100	200	300	400	500	600	700
120 Volts	Wire gage number								
¼	12	12	10	8	6	4	4	4	3
⅓	12	10	10	6	4	4	3	2	1
240 Volts									
¼	12	12	12	10	8	6	6	4	4
⅓	12	12	12	10	8	6	6	4	4
½	12	12	12	10	8	6	6	4	4
¾	12	12	10	8	6	4	4	4	3
1	12	10	10	6	4	4	3	2	1
1½	12	10	10	6	4	4	3	2	1
2	10	8	8	4	3	2	1	0	0
3	10	8	6	4	2	1	0	00	000
5	8	6	4	2	0	00	000	4/0	4/0

(If overhead wiring is used and span is more than 50 feet—use 8-gage wire for strength; or larger if needed for motor. Spans under 50 feet can use 10-gage wire unless larger size is needed for motor.)

TYPE OF CONDUCTOR NEEDED

There is a wide variety of coverings for single wires and for cables, which have been developed for both copper and aluminum wire to meet a wide range of operating conditions. The American Fire Protection Association, through its publication the *National Electrical Code,* has identified these different types by letters such as TW, UF, and so forth.

The **type of wiring you use will be determined by the conditions you have** in extending electrical service from your power source to the point where you will connect to your pump. For example, your wiring may amount to (a) extending the wiring *inside an existing building,* or (b) using *overhead wires* to extend from a building or pole to the pump location, or (c) extending the wiring *underground* from a pole or building to the pump. Whatever the condition, there are two or more types of wiring to fit your needs.

If the wiring to your pump will be **inside of a building** where it will be supplied from existing wiring, the cable types most readily available will be the following:

Type NM - A nonmetallic sheathed cable. The outer sheath covering over the insulated wires is moisture resistant and flame retardant. It may have an uninsulated (bare) wire included in the sheath for grounding purposes.

It can be installed in both exposed and concealed work in locations that are normally dry. It can be used in frame or concrete structures where there is not excessive moisture or dampness.

Type NMC - Similar to NM except it can be used in damp and corrosive locations.

For **overhead wiring** you can use either single wire conductors or cable. The types normally available for this purpose are the following:

Single-Wire Conductors

Type TW - A conductor with a thermoplastic covering which is both flame resistant and moisture resistant. It can be used in wet or dry locations.

Type THW - A conductor with a thermoplastic covering which is a flame retardant. It provides both moisture and heat resistance and can be used in wet or dry locations.

Cable

Type SE - A cable with a flame-retardant moisture-resistant covering made up of two or more conductors, one of which may be without insulation.

If you plan to extend your **wiring underground** from your meter service to your pump, you will be using cable. The types of cable normally available for underground use are the following:

Type UF - A cable with a tough plastic covering which is flame retardant, moisture resistant and suitable for direct burial in the ground (Figure 134).

It should not be used where it is exposed to the direct rays of the sun. The covering tends to break down in direct sunlight.

Type USE - A service-entrance cable that can be used for burial in the ground. It has a moisture-resistant covering.

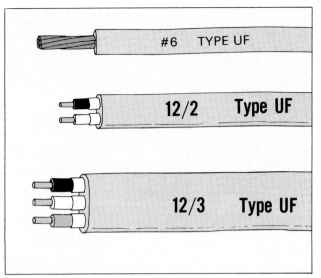

FIGURE 134. The UF cable can be placed underground. It is often laid in the trench with the pipe. Also satisfactory for interior use.

C. What Electrical Protection to Provide

Once you have selected the proper wire size and the proper type of wire for your needs, your next decision is involved in protecting the motor on your pump so it will give uninterrupted service. There are two ways a motor can be damaged so that it will either not work satisfactorily or will stop completely. One is for the motor to become **overloaded** causing the windings to burn out, or for a **lightning surge** to enter the motor and break down the insulation on the windings. You can protect your motor from both of these conditions with equipment now available and with the proper installation.

The information you need for providing such protection is discussed under the following headings:

— Feeder circuit protection
— Pump motor protection

FEEDER CIRCUIT PROTECTION

No matter how small your water system, it can be of help when a fire first starts. But to be of help, you should make certain it will continue to operate—even though the rest of the circuits are shorted or burned out during the fire. For that reason the pump should be

1. Placed on an entirely separate circuit from other equipment.
2. Connected as closely as possible to the point where your meter is located.

If your water system is in the same building with the meter, run the circuit from the service entrance through the basement or under the building if you have no basement. Fire usually burns out the upper part of a structure first. This arrangement may enable your pump to keep running long after the other circuits are burned out.

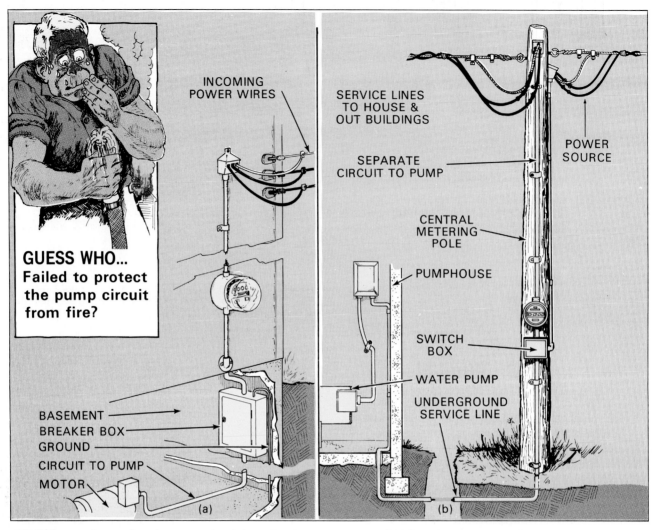

FIGURE 135. Protecting the pump circuit from fire damage. (a) Provide a separate circuit for the pump motor and keep the circuit as short as possible to help avoid fire exposure. (b) With a central metering pole, extend the pump circuit underground from the pole to the pump for maximum protection against burnout.

An even better arrangement is that shown in Figure 135a if you can arrange it. By keeping the distance from the service-entrance fuse box to the pump as short as possible, you may further reduce the chance of a fire shorting out the pump circuit.

If your water system is in a structure of its own, you may be able to extend a circuit underground from the service-entrance location to your pump. The underground wires are completely protected from fire.

If you have a **central metering pole,** a still better arrangement is to extend your pump circuit directly from it to your pump (Figure 135b).

PUMP MOTOR PROTECTION

Your pump motor needs two kinds of protection: (1) overload protection and (2) protection from lightning surges.

Overload protection is provided to keep the pump motor windings from burning out in case of extra-heavy loads—loads that are more than the motor is designed to handle. This can happen, for example, with a centrifugal pump if it is pumping at full capacity and with free discharge of water when it is designed to pump against pressure.

To protect the motor from overloads of this type, it should have a separate overload protection device. There are different types—**circuit breakers, time-delay fuses and magnetic switches**—with overload protection. The latter are commonly used with motors of 1 horsepower or larger. Your dealer will help you

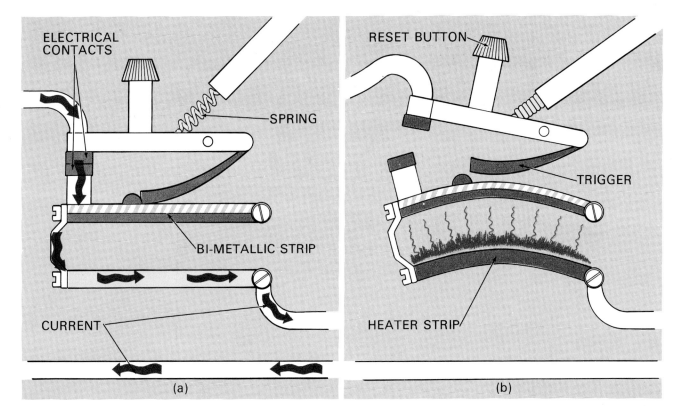

FIGURE 136. How a manual-reset, bimetallic overload control works. (a) Animated movement of electrons (current) through the heater strip when operating within its rated capacity. (b) With an overload, the heater strip becomes hot. The heat, being so close to the bimetallic strip, causes it to heat and bend. When heated sufficiently, the bimetallic strip bends enough to release the trigger. The spring pulls the electrical contacts apart, thus opening the circuit. After the bimetallic strip has cooled, it straightens to its original position as in (a). Then, you can press the reset button and reset the breaker which turns on the current.

select the kind that best fits your needs. Figure 136 shows how one type of overload control works.*

Interest in **protection from lightning surges** has developed since submersible pumps have become popular. It was assumed at one time that motors on submersible pumps would be more likely to be damaged by lightning surges than motors on other kinds of pumps. It developed that this was not true, but there was one other factor. When damage did occur, the cost of repair was usually higher than for other water systems. Consequently, protection against lightning surges has come to be accepted as good insurance. Most submersibles have internal protection built in the motors for added protection (Figure 137).

Lightning surges develop on power lines during a thunderstorm. The power lines are seldom hit directly with a bolt of lightning, but the lightning action induces electrical surges in the lines when there is lightning action in the vicinity. The high-voltage surges coming in on the motor circuit from the power line can damage the motor by puncturing the insulation on the motor windings. This permits the regular line current to leak through the insulation and cause a short. The windings then overheat, and the insulation is damaged.

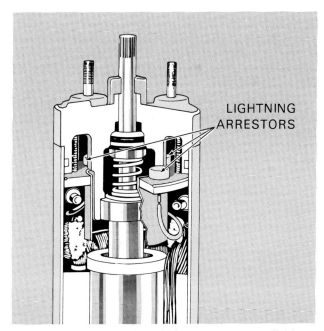

FIGURE 137. Built-in lightning arresters are available on most motors for submersible pumps.

---

*Details regarding types of overload controls for motors and how they work are given in the publication "Electric Motors—Selection, Protection, Drives." American Association for Vocational Instructional Materials, Engineering Center, Athens, Georgia 30602.

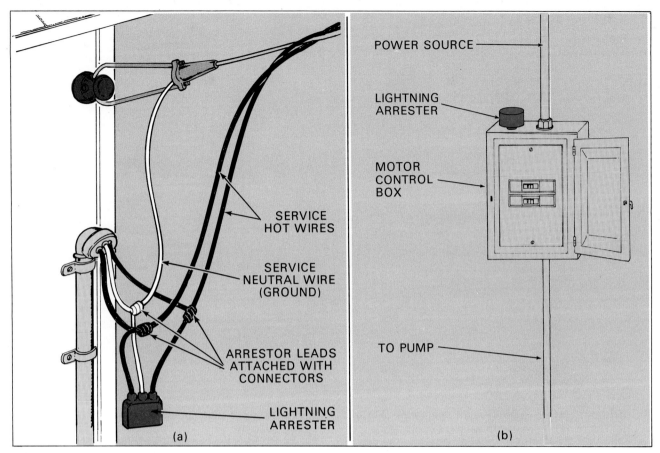

FIGURE 138. A pump motor can be protected from lightning surges by (a) a lightning arrester connected to the service-entrance conductor, or (b) a lightning arrester mounted in the switchbox that serves the pump motor.

To protect against this situation, you can install a **lightning arrester** at the point where your *service wires are connected to the service entrance cable*. The arrangement shown on Figure 138a protects not only your pump motor, but your whole wiring system from lightning surge damage.

Another option is to install the lightning arrester in the **motor-control box** (Figure 138b) and protect the motor only. With either arrangement, the grounding wire from the lightning arrester must have an excellent contact with moist earth. This may be a ground rod or metal well casing, or a metal drop pipe that extends from the pump into the well.

With a satisfactory ground, when a lightning surge develops that approaches being strong enough to puncture the motor insulation, it passes instead through the lightning arrester directly through the ground connection to ground so that no damage is done.

Figure 139 shows a **multi-ground arrangement.** Note that a driven ground rod is used at the building, a bare stranded copper wire extends from that ground to the well casing where it is strapped tightly to the casing. Both of these are excellent grounds. The wire then extends into the well and to the pump motor where it is bonded to the motor frame. Since the submersible pump is in water, this is certain to provide an excellent ground.

If a metal pipeline is used between the pump and the residence, the ground wire is also bonded to it. If plastic pipe is used, bonding is not necessary, because plastic pipe is a non-conductor.

With all of these ground connections, you will have an excellent low-resistance means by which lightning surges can be conducted into the ground.

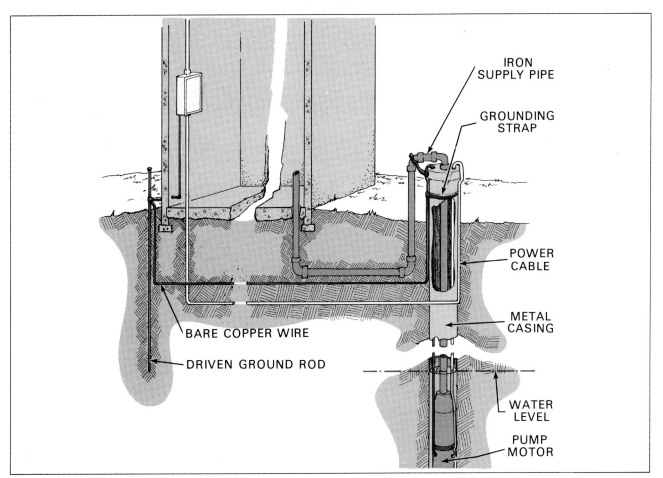

FIGURE 139. Suggested method of providing several ground connections to assure easy passage of a lightning surge into the ground.

Notes

Acknowledgments

Acknowledgment is given for the excellent suggestions, criticisms and assistance given by the following in the preparation of the publication:

EDUCATION

Alberta

Taylor, Douglas A., Research and Development Specialist, Alberta Agricultural Extension, Canada.

Arizona

Jacobs, Clinton O., Professor, Agricultural Education, University of Arizona.

California

Weicker, Theodore, 177 Bolivar Lane, Portola Valley, CA.

Georgia

Allison, James M., Associate Professor; **Brown, Robert H.,** Professor and Division Chairman and **McLendon, B. Derrell,** Associate Professor, Agricultural Engineering, University of Georgia.

Colvin, Thomas S., Research and Development Specialist, AAVIM; **Winsett, Ivan L.,** Executive Director, Georgia Electrification Council.

Samples, Lawton, Extension Agricultural Engineer, Tifton, GA.

Illinois

Anderson, Don, Coordinator, Water Supply Operations Training, Environmental Resources Training Center, Southern Illinois University at Edwardsville.

Iowa

Hoerner, Thomas A., Professor, Agricultural Engineering, The Iowa State University.

Kansas

Holmes, Elwyn S., Extension Agricultural Engineer, Kansas State University and **Slyter, Damon E.,** Agricultural Education Program Specialist, Kansas State Department of Education.

North Carolina

Kane, Barney, Department of Environmental Health, East Carolina University, Greenville, NC.

Tart, C. V., Chief Consultant, Agricultural Education, NC.

Texas

Buchannan, J. B., Stinnette, TX.

Wisconsin

Doering, F. J., Supervisor, Wisconsin Agricultural Education.

INDUSTRY

Aqua Fire Energy Corporation (Ferr-X Division)

Hament, Louis, P.O. Box 356, Clifton N.J. 07011

Autotrol Corporation, Land Products Inc.

Grout, Edward, General Manager, and **Hament, Louis,** 1701 West Civic Drive, Milwaukee, WI.

Burks Pumps

Haelich, Bill, P.O. Box 431, Decatur, IL 62525.

Campbell Enterprises

Campbell, Gordon M., Willits, CA.

Decatur Pump Company

Decatur, Illinois

Fairbanks-Morse

McCormick, Barry, National Service; **Ken Priebe,** Senior Buyer and **Gene Strautman,** National Account Manager, Fairbanks-Morse—Pump Division, 3601 Fairbanks Ave., Kansas City, KS 66110.

Filtrine Manufacturing Company, Waldwick, N.J.

Gould, Inc.

Will, L. A., Marketing Service, Electric Motor Division, St. Louis, MO.

Jacuzzi
Ford, Jim, 11511 New Brenton Highway, Little Rock, AR 72203.

Land Products, Inc.
Grout, Edward C., General Manager, Milwaukee, WI.

Lakos Separators
Delenikos, Randy, 1911 N. Helm, Fresno, CA 93703

Claude Laval Corporation
Galanica, Randy.

F. E. Meyers Co.
Braun, J. L., Product Manager, Water Systems, Ashland, OH.

Red Jacket Pumps
Green, Ron, 500 E 9th Street, Davenport, IA 52807.

Rife Hydraulic Engine Mfg. Co.
Gibbs, Jack, Vice President, P.O. Box 415, Andover, N.J. 07821

Sta-Rite Industries, Inc.
Atkins, Jim, Director of Training, Sta-Rite Training Institute, Delavan, WI 53115.

The Lindsay Company
Smith, Floyd R., Merchandising Manager, P.O. Box 43420, St. Paul, MN 55164.

Tait Manufacturing Company, Dayton, Ohio

The Valley Pump Group
Wegehoft, Dick, Box 1364, Conway, AR 72032

Universal Oil Products Company, Figure 16 reprinted with permission from "Ground Water and Wells," copyright 1966.

GOVERNMENT AGENCIES

Department of Water Resources
Riley, Al, Resources Planner, Sacramento, CA.

SCS-USDA
Martin, Cecil, Civil Engineer, Athens, GA, **Twitty, Ken,** Agricultural Engineer, Ft. Worth, TX.

U.S. Environmental Protection Agency
Fox, Kim R., and **Lauch, Richard P.,** Engineers, Inorganic Contaminant Removal Research; **Geldriech, Edwin E.,** Microbiologist; **Logsden, Gary S.,** Engineer, Turbidity and Microorganism Removal; **Love, O. Thomas,** Engineer, Organic Contaminant Removal Research and **Sorg, Thomas J.,** Engineer, Inorganic Contaminant Removal, Point of Use.

OTHERS

National Water Well Association
Lehr, Jay H., Executive Director, **Poehlman, Jim** and **Smith, Stuart,** Worthington, OH.

Texas Highways Magazine, Austin, Texas

Water Systems Council
Lane, Joe, Chairman, Tank Committee and **Williams, Pamela,** Associate Director, Chicago, IL.

References

SPECIFIC REFERENCES

1. *Feeds and Feeding;* Morrison, Frank B.; Morrison Publishing Co.

2. *New Concepts in Farmstead Water-System Design;* Jones, Elmer E.; Transactions of the American Society of Agricultural Engineers, Vol. 11, No. 3, 1968.

3. *Response of Hens Under Thermal Stress to Dehydration and Chilled Drinking Water;* Wilson, W. L. and Edwards, W. H.; Div. of Poultry Husbandry, Univ. of Calif., Davis, Calif.

4. Goff, O. E., and Littlefield; Univ. of Tenn., Knoxville, Tenn. unpublished data.

5. *Cooling and Ventilating the Laying House;* Hobgood, Price and Jaska, R. C.; Texas A & M College, College Station, Texas.

6. *Cooling Poultry Houses in the Southeast;* Drury, Liston N.; Ag. Research Service; USDA, Univ. of Ga. February 1961 (mimeo).

7. *Evaporative Cooling of Caged Laying Hens;* Hart, S. A., Woodard, A. E. and Wilson, W. O.; Univ. of Calif., Davis, Calif., June 1961 (mimeo).

8. Report of the Chief of the Bureau of Plant Industry, Soils, and Agricultural Engineering; Ag. Research Administration, 1951.

9. *Your Pigs May Need More Water;* Altman, Landy B., Ashton, Gordon C. and Catron, Damon; Iowa Ag. Exp. Sta., Iowa Farm Science FS-446.

10. *Pamper Your Pigs with Wallows;* Self, H. L.; The Progressive Farmer, Aug. 1954.

11. *Mist Cooling Increases Swine Gains;* Research Report, Vol. 3 No. 2, Ag. Exp. Sta., Univ. of Fla., Gainesville, Fla.

12. *1960 Annual Report;* Univ. of Ga., College of Agriculture Exp. Stations.

13. *Cool Hogs Do Better;* Clawson, A. J.; Research & Farming, Vol. XV: No. 4.

14. *The Use of Sprinklers and Wallows for Cooling Swine;* Andrews, F. N., Culver, A. A., Noffsinger, T. L. and Fontaine, W. E.; Purdue Univ., Ag. Exp. Sta., Mimeo. A. H. 186.

15. *Effects of Modified Summer Environment on Swine Performance;* Heitman, Hubert, Jr., Bond, T. E., Kelly, C. F. and Hahn, LeRoy; Univ. of Calif. and USDA, November 1959.

16. Oklahoma Station Cir. #C131, out of print.

17. *Farmstead Water Demands and Peak Use Rates;* Yung, F. D.; Univ. of Neb., Ag. Eng.; Paper No. 61-212, June 1961.

18. *Summary Report on the Residential Water Use Research Project;* Linaweaver, F. P. Jr., Geyer, John C. and Wolff, Jerome B.; Journal, American Water Works Association, Vol. 59, March 1967.

19. *Dairy Farmstead Water Use;* Jones, Elmer E.; Agricultural Engineer, Agricultural Research Service; Paper No. 64-232, American Society of Agricultural Engineers, June 1964.

20. *Evaluation of the Tennessee Water Supply Program;* Bureau of Water Hygiene, Environmental Protection Agency, Region IV, January 1971.

21. *Manual of Individual Water Supply Systems;* Public Health Service Publication No. 24; U. S. Dept. of Health, Education and Welfare, Revised 1973.

22. *Operation of an Electric Type Diaphragm Pump Chlorinator;* Hill, Ronald D.; Ohio Ag. Exp. Sta., March 1961 (mimeo).

23. *Retention of Water for Disinfection in Chlorinated Small Water Supplies;* Baumann, E. R. and Ludwig, D. D.; Iowa State Univ., Ames, Iowa; Engineering Report 36, 1962.

24. *Measuring Kill Rates of Chlor-Dechlor Systems;* Renn, Charles E., Barada, M. F. and Christian, E. D.; Journal of the American Society of Agricultural Engineers, October 1965.

25. *Radioactive Fallout in Time of Emergency, Effects Upon Agriculture;* Ag. Research Service, USDA, April 1960.

26. United States Environmental Protection Agency; Safe Drinking Water Act.

27. *Identification and Removal of Herbicides and Pesticides;* Sigworth, E. A.; Journal American Water Works Assn., Aug. 1965.

28. *Better Tools for Treatment;* Black, A. P.; Journal American Water Works Assn., Feb. 1966.

29. *Quality of Water in Ohio Farm Ponds;* Hill, R. D., Schwab, G. O. Malaney, G. W. and Weiser, H. H.; Research Bul. 922, Ohio Ag. Exp. Sta., Oct. 1962.

30. *Cultivation, Morphology, and Classification of Iron Bacteria;* Wolfe, Ralph S.; American Water Works Assn. Journal, Sept. 1958.

31. *Water Systems for Rural Fire Protection 1969;* National Fire Protection Assn.

32. *Intermediate Storage for Farmstead Water Systems;* Jones, Elmer E.; Agricultural Engineering Research Division, ARS, USDA; issued by the Cooperative Extension Service, Univ. of Maryland, March 1964.

33. *What About Water After a Nuclear Attack?;* Federal Extension Service, USDA, Fact Sheet No. 12.

34. *Wanted: Higher Pressure Settings for Greater Pump Potential;* Journal Plumbing—Piping Hydronics, July 1969.

35. *Fire Protection Potential of Individual Water Systems;* Jones, Elmer E.; Agricultural Engineer, USDA; October 1964 (mimeo).

36. *Protect Water Pipes with Electric Heat Cable;* McFate, Kenneth L.; Agricultural Engineer, Missouri Cooperative Extension Service; October 1965.

GENERAL REFERENCES

A Drop to Drink—A Report on the Quality of Our Drinking Water. United States Environmental Protection Agency, June, 1976.

A System for Pond Water Purification. Science and Technology Guide, University of Missouri—Columbia Extension Division, January, 1972.

Agriculture and Clean Water. Midwest Research Institute, August, 1977.

An Alternative Method of Iron Removal. Water Well Journal, August, 1978.

Annual Report 1979. United States Department of the Interior, Office of Water Research and Technology, United States Government Printing Office, 1980.

Bacteria—What Are They? Water Well Journal, February, 1980.

Basic Gas Chlorination Manual, Training and Licensing Section, Ministry of Environment. Toronto, Ontario, Canada, 1972.

Basic Water Problems. Water Well Journal, June, 1976.

CISTERN WATER—Its Protection and Treatment, Brooks, Jesse B. Cooperative Extension Service, University of Kentucky, June, 1974.

Captured Rainfall, Small Scale Water Supply Systems, Johnson, Huey D., Brown, Edmund G. Jr., Robie, Ronald B., State of California, The Resources Agency, Department of Water Resources, May, 1981.

Chemical Contamination. Special Legislative Commission on Water Supply/Commonwealth of Massachusetts, October, 1981.

Clean Water—Report to Congress 1975-76. United States Environmental Protection Agency, United States Government Printing Office, 1977.

Clearing the Waters on Home Filters. Consumer's Digest, January-February, 1980.

Codes and Piping. DE Journal, March, 1976.

Cold Weather Protection of Water Systems in Mobile Homes. Extension Service, Pennsylvania State University.

Conservation Practices Help Meet Clean Water Goals. Soil Conservation, August, 1978.

Continuous Chlorination of Contaminated Water. Water Well Journal, October, 1977.

Corrosion in Oil and Gas Production. Jorda, R. M., Production Technology Save Series 1231: Shell Oil Company, 1973.

Drinking Water and Human Health: Part I. Water Well Journal, March, 1978.

Directory of Manufacturers Directory of Suppliers, Water Well Journal, January, 1982.

Domestic Wastewater Treatment & Disposal. State of the Art: Water Well Journal, June, 1979.

EPA Standards for Drinking Water Become Effective in June 1977. The Johnson Drillers Journal, March-April, 1976.

Focus on Water—Who Gets the Rights to It?, Harl, Neil E.

Ground Water Pollution in the South Central States, Scalf, M. R., Keeley, J. W., and LaFevers, C. J., National Environmental Research Center, Environmental Protection Technology Series, United States Environmental Protection Agency, June, 1973.

Home Filters to "Purify" Water. Changing Times, February, 1981.

How to Keep Your Drinking Water Safe, Schultz, Mort, Popular Mechanics, March, 1975.

How to Install a Water Conditioner. Water Well Journal, June, 1976.

Hydrogen Sulfide Problems. Water Well Journal, December, 1977.

Improving Nature's Wonder—Water. Lindsay Division, St. Paul, Minnesota.

Individual Water Program. State of Tennessee, Department of Public Health, Nashville.

It's All on the Nameplate: Everything You Always Wanted to Know About Jet Pumps. Water Well Journal, March 1977.

Lightning, The Enforcer, Myers, Co., F. E.

Manual of Individual Water Supply Systems. Environmental Protection Agency, Water Supply Division, January, 1974.

Mercury Pollution Control in Stream and Lake Sediments. United States Environmental Protection Agency, March, 1972.

National Interim Primary Drinking Water Regulations. Environmental Protection Agency, Office of Water Supply, United States Government Printing Office, 1977.

Private Water Systems Handbook. Midwest Plan Service, Iowa State University, 1979.

Protecting Ground Water From Domestic Wastewater Effluent. Water Well Journal, June, 1977.

Pumping Water by the Air-Lift Method Has Practical Applications. The Johnson Drillers Journal, November-December, 1973.

PVC Water Well Casing: The First Hand Experiences of Two Drillers. Water Well Journal, December, 1975.

Removing Iron From a Water Supply. Water Well Journal, October, 1977.

Reverse Osmosis Process Solves Water Treatment Problems. Reported from Paper No. 73-2502, American Society of Agricultural Engineers, May, 1974.

Safe Drinking Water for Rural America, Summers, C. L., Lee, W. A. Nebraska State Department of Health, Division of Environmental Engineering, June, 1976.

Scale in Oil Field Water Handling Facilities, Production Technology Save Series 1191: Jorda, R. M. Shell Oil Company, 1972.

Sizing Plastic Pipelines for Water on the Range. Cooperative Extension Service, New Mexico State University, April, 1975.

The Impact of Hazardous Wastes on Ground Water: Part I. Water Well Journal, October, 1978.

The Nation's Water Resources 1975-2000, Volume I: Summary. United States Water Resources Council, December, 1978.

The Public Benefits of Cleaned Water: Emerging Greenway Opportunities. United States Environmental Protection Agency, Office of Land Use Coordination, August, 1977.

Treatment at the Tap—A Homeowner's Guide to Water Filters. Rodale's New Shelter, April, 1980.

Water Conditioning—Bacteriological Safety. Water Well Journal, October, 1976.

Water Conditioning—Quality Problems: Turbidity, Color, Odor & Taste. Water Well Journal, November, 1975.

Water Conditioning—Reverse Osmosis. Water Well Journal, July, 1976.

Water: Pinch on Energy Development. Electric Power Research Institute Journal, October, 1979.

Water Research Capsule Report: Water Factory 21. Office of Water Research and Technology, United States Environmental Protection Agency, United States Government Printing Office, 1978.

Water Sampling Made Easier With New Device. The Johnson Drillers Journal, July-August, 1976.

Water-Saving Devices: The State of the Art, Sharpe, William E., Fletcher, Peter W.; DE Journal, October, 1975.

Water Supply Disinfection With Iodine and Silver. Water Well Journal, February, 1978.

Water Systems Handbook. 7th Edition, Water Systems Council, Chicago, Illinois.

Water System Operator Needs Basic Information to Keep Pumps Working. The Johnson Drillers Journal, March-April, 1974.

Water Treatment: Part Ten—Water Analysis. DE Journal, October, 1972.

What's In the Water? Water Well Journal, September, 1978.

What You Should Know About Superchlorination—Dechlorination. Water Well Journal, June, 1976.

Metric Conversion Tables

MEASUREMENT CONVERSION FACTORS (APPROXIMATE)

Metric		× Conversion Factor	=	Customary
LENGTH				
mm	millimeters	0.04	inches	in
cm	centimeters	0.4	inches	in
m	meters	3.3	feet	ft
m	meters	1.1	yards	yd
km	kilometers	0.6	miles	mi
AREA				
cm²	square centimeters	0.16	square inches	in²
m²	square meters	1.2	square yards	yd²
km²	square kilometers	0.4	square miles	mi²
ha	hectares (10,000 m²)	2.5	acres	
MASS (weight)				
g	grams	0.035	ounces	oz
kg	kilograms	2.2	pounds	lb
t	tonnes (1000 kg)	1.1	short tons	
VOLUME				
ml	milliliters	0.03	fluid ounces	fl oz
l	liters	2.1	pints	pt
l	liters	1.06	quarts	qt
l	liters	0.26	gallons	gal
m³	cubic meters	35.3	cubic feet	ft³
m³	cubic meters	1.3	cubic yards	yd³

Customary		× Conversion Factor	=	Metric
LENGTH				
in	inches	2.54	centimeters	cm
ft	feet	30.5	centimeters	cm
yd	yards	0.9	meters	m
mi	miles	1.6	kilometers	km
AREA				
in²	square inches	6.5	square centimeters	cm²
ft²	square feet	0.09	square meters	m²
yd²	square yards	0.8	square meters	m²
mi²	square miles	2.6	square kilometers	km²
a	acres	0.4	hectares	ha
MASS (weight)				
oz	ounces	28	grams	g
lb	pounds	0.45	kilograms	kg
	short tons (2000 lb)	0.9	tonnes	t
VOLUME				
tsp	teaspoons	5	milliliters	ml
tbsp	tablespoons	15	milliliters	ml
fl oz	fluid ounces	30	milliliters	ml
c	cups	0.24	liters	l
pt	pints	0.47	liters	l
qt	quarts	0.95	liters	l
gal	gallons	3.8	liters	l
ft³	cubic feet	0.03	cubic meters	m³
yd³	cubic yards	0.76	cubic meters	m³
FORCE/AREA				
psi	pound force/in²	6.89	kilo paschals	kPa
psi	pound force/in²	6.89	newton/meter²	N/m²
psi	pound force/in²	.070	kilogram/meter²	Kg/m²
psi	pound force/in²	.069	bar	bar

Notes

Index

A

Absorption barrier float............107
Acid/chemical treatment............22
Acidity......................54, 55
Acid..........................44, 64
Acids............................51
Acid Water....................55, 64
Activated carbon..................66
Activation........................66
Air-handling.....................86
Air-release unit.................111
Air-volume control...............107
Algae...................55, 66, 67, 68
Alkaline.........................44
Alum feeder...................68, 69
Alum solution....................69
Ammonia..........................43
Anthrax..........................25
Aspirator........................38
Atmospheric pressure.............81
Automatic drain..................30
Automatic recharge...............61
Automatic-roofwash............30, 31
Available chlorine...............36

B

Backwashing..........22, 63, 65, 67, 71
Bacteria,..........25, 41, 42, 43, 46, 58
   coliform......................33
   iron................54, 62, 63, 64
   manganese......................54
Bailer...........................16
Bituminous coal..................66
Bleeder valve..........108, 110, 111
Brucellosis......................25

C

Cable
   NM nonmetallic sheathed........142
   NMC...........................142
   SE............................142
   service entrance..............146
   UF............................143
   USE...........................143
Calcium................52, 58, 59, 60, 63
Calcium hypochlorite..........36, 47
Calibrated.......................19
Carbon dioxide............51, 52, 55
   free..........................54
Catchment Area................20, 21
Caustic soda..................64, 65
Cells
   elastic.......................114
   elastic storage...............113
   photo-electric.................42

Central metering pole............144
Centrifugal pump............20, 83, 84
                  85, 86, 96, 107, 112
Centrifugal-jet (ejector) pump....85
Centrifugal-submersible pump......86
Cesspool......................27, 33
Chemical feeder...............62, 64
Chemical sealant.................55
Chemical waste...................55
Chlorine.................35, 36, 38, 43
          44, 45, 47, 49, 55, 62, 64
Chlorine demand..................43
Chlorine disinfection............44
Chlorine gas.....................36
Chlorine residual.........43, 44, 46
Chlorine solution........46, 47, 49, 65
Chlorine supply..................36
Chlorine taste............44, 47, 65
Chlorination.....................44
   shock.........................22
   simple..................44, 45, 47
   super..................43, 46, 47
Chlorinator...................49, 64
   injector type.............36, 38
   pump type.....................36
   tablet type..............36, 39
Chlorinator units................36
Combined available chlorine......49
Combined chlorine residual.......44
Circuits
   feeder...................141, 143
   pump.........................144
   120-volt.....................141
   240-volt.....................141
Circuit breakers................145
Cisterns.........13, 15, 20, 21, 22, 23
                26, 32, 31, 33
Cistern capacity.................23
Cistern construction.............31
Cleaning agents..................36
Cleaning solvents................66
Closed-type single acting
   cylinder......................88
Coagulation..............67, 68, 70
Commercial laundry bleach........36
Competing sustained use..........75
Compounds
   polymer.......................69
   polyphosphate.................62
Conductor..................141, 142
   aluminum.....................141
   copper.......................141
   THW single-wire..............142
   TW single-wire...............142
Contact time.............44, 46, 47

Contaminants.....................25
Contamination....................34
   underground...................33
Copper sulfate...................68
Copper tubing....................18
Corrosion........................51
Corrosive........................47
Cut-off valves........135, 127, 132, 136

D

Demand allowance........133, 134, 135
Diaphragm.................36, 37, 108
Diaphragm-type air control......108
Diatoms..........................70
Diatomite-filter material........70
Diffuser.........................85
Disinfectant.....................49
Disinfecting unit.............43, 47
Disfection unit..................48
Dissolved iron...................54
Distillers.......................57
Diversion ditch..................30
Diversion valve..................30
Dosage...........................43
Double-acting deep well cylinder....88
Double acting principle..........82
Drain port......................138
Drawdown............16, 17, 18, 19, 20
Drawdown control switch.........106

E

Earth auger......................14
Efficiency.......................96
Elastic pressure cell...........104
Electric heating tape...........137
Electricity.....................139
Environmental Protection Agency....26
Eureka cylinder..................88

F

Farm animals.....................11
Farm needs....................69, 73
Filter
   activated carbon..............59
   bed-type activated carbon.....67
   carbon bed-type activated
      carbon.....................66
   precoat cartridge type activated
      carbon.....................66
   ceramic cylinder..............67
   Diatomite.................67, 70
   Diatomaceous earth............70
   iron..........................63
   iron and sulfur..............124
   makeshift.....................67

155

neutralizing 124
oxidizing 63
   Iron removal 63
pads 67
paper 67
pre-coat 67
sand 30, 31
   rapid sand 67, 70
   slow sand 69
Filter aid 71
Filtering element 70
Filtration 32
Fire protection 11
Fixture-flow rate 124
Flame-retardant 142
Float switch 106
Floc 69
Fluorescin 34
Food-grade phosphate 62
Foot-valve 112, 47
Foot-valve assembly 47
Free available chlorine 49
Free chlorine residual 44

G

Gage
   pressure 18, 113
   vacuum 19
Gardening use 75, 78
Garden watering 122
General contamination 49
Giardia cysts 25
Gravity systems 139
Gravity tank, ... 46, 105, 106
   covered 104
Gravity feed tanks 99
Ground water source 29

H

Hardpan 26
Hatch 119
Heat 35
Hepatitis 25
Home water uses 11, 69, 73, 75, 78
Horsepower 141
Hydrant 123
   frost-proof 138
   outside 123, 124
   yard 124
Hydraulic rams 139, 140
Hydropneumatic tank 131
Hypochlorites 36
   sodium 36

I

Impeller 83
Inset 38
Installations
   deep-well 87
   pit-type 116
Intake valve 37
Intermediate storage tank 101
Intermediate storage 103
Intermittent use 75
Internal combustion engines 140
Ions 59
   sodium 59
Ion exchange 63
Ion exchange unit 59
Ion exchange zeolite unit 63
Iron 35, 62, 63, 64
Iron particles 54, 63
Iron (dissolved) 62
Irrigation 76

J

Jet 38, 85

Jetting
   high-velocity 22
Jet pumps 85, 88, 96, 107, 112
   deep well 86, 96, 108, 113
   shallow well 86

L

Lake 13, 15, 22, 25, 32, 50
Laundry bleach, ... 47
   domestic 36
Lawn use 75, 78
Leakage 23
Lightning arrester 146
Lightning surge 143, 145
Lignite 66
Limestone 65
Livestock use 73, 75, 76, 78

M

Magnesium 52, 58, 59, 60, 63
Magnetic switches 145
Maintenance 62, 63, 64, 65, 67, 69
                                70, 71
Manganese 54, 63, 64
Manganese treated green sand 63
Marble chips 65
Mechanical damage 136
Mechanical servicing 49
Mercury 19
Micro-organisms 26
Micromet 62
Minerals 51
Mixing tank 47
Moisture-resistant 142
Motor-control box 146
Multi-ground arrangement 146
Multi-stage pumps 84

N

Nalco M-1 62
National Interim Primary Drinking
   Water Regulation 32
NSF—National Sanitation Foundation
   Testing Laboratories, Inc. 130
Neutralizer 65
Non-competing sustained use 75

O

Offset 95
Open gravity tank 104
Open water supplies 50
Operating principle 40
Organic compounds 66
Organic material 67
Organic matter 54, 55
Orthotolidine 49
Osmosis 58
   Reverse 58, 59
Outlets
   water outlets 46, 122, 138
   outside 121
   water-use 121, 122, 136, 139
Overhead wires 142
Overload 143
Overload protection 144
Overpumping 22
Oxidation 55
Ozonators 57
Ozone treatment 35

P

Parasites 25
Peak demand 73, 75, 79, 123, 124, 125
Pesticides 66
Petroleum coke 66
Phosphate 62
Phosphate feeders 62
Pipe connections 131

Pipe
   copper 46, 128, 133, 135
   type "K" copper pipe 128
   type "L" copper pipe 128
   inlet 31
   outlet 31
   polybutylene pipe 133
   steel 133, 135
      galvanized steel 128
   suction 80, 81, 83
   waste 31
Plain steel tank with floating
   wafer 98
Plastic pipe 46, 128, 133, 135, 138
   PE 128, 130
   PVC 130
Plastic tubing 18
Polio (poliomyelitis) 25
Pollutants 50
Pollution 13, 25, 26, 27, 28
                     30, 31, 33, 104, 138
   well 116
Pond 13, 15, 22, 23, 25
                         26, 32, 50, 67
Potassium permanganate 63
Powdered gypsum 68
Power source 139
PPM (parts per million) 35, 43
                             44, 46
Precharged Diaphragm/Bladder
   Tank 97
Pressure adjusting screw 112
Pressure tank 104
Pressure storage tanks 97
Pressure switches 105
Privy 27
Privies 33
Psi 94, 95, 106, 130, 136
Pump efficiency 96
Pump and motor combination 94
Pumping capacity 136
Pumps
   deep well centrifugal 84
   shallow well centrifugal 84
   piston 81, 112
   deep well piston 82, 87, 107, 112
   shallow well piston 82, 107, 112
   positive-acting 96
   deep well 81, 94
   shallow well 81, 87, 94, 96
   submersible 96
   test 16, 20
   turbine 86, 107
   deep well turbine 86
   shallow well turbine 86, 114
   water 78
PVC piping material 128

Q

Quartz sleeve 41, 42
Quartz tube 42

R

Radioactive dust 49, 50, 102
Radioactive fallout 25, 49, 50, 102
Rainfall 21
Rainwater 20, 30, 51
Rawhiding 22
Recharging 61, 63
Redevelopment 22, 23
Reservoirs, ... 15, 22, 46, 67, 70
                     99, 104, 105, 106, 140
   pumped storage 99
Residual 43, 44, 46
Restricting valve 39
Retention time 44, 46
Retention tube 40

Retrench 136
Rust particles 54, 62, 63

S

Sanitary well seal 28
Saturated 67
Scale 51
Sediment 70, 71
Sedimentation 67, 68, 69, 70
Sediment trap 114
Septic tank 27
Service entrance 144
Service entrance fuse box 144
Shallow trenches 138
Shallow well float type control ... 107
Sill cock 122
Single-phase motors 142
Slime 63
Sludge 47
Snifter valve 108, 110, 111, 116
Soda ash 64, 65
Sodium 60
Sodium bicarbonate 64
Solenoid valve 40
Solution 47
Solution container 47
Springs, ... 13, 15, 20, 22, 23
                     26, 30, 33, 34, 50
  surface-flowing 22
Sprinkling
  Garden 11, 12, 124
  Lawn 11, 12, 124
Stop-and-waste cock 137
Stop-and-waste valve 136
Sulfur 35
Sump 116
Surging 22
Sustained use 75

T

Tank
  neutralizing 65
  settling 68, 69
  plain steel 98
  storage 70, 140
Tarpaulin 50
Taste
  bitter 66
  brackish 66
  chlorine 66
  fishy 66
  marshy 66
  metallic 66
  oily 66
  rotten egg odor and taste 66
  salty 66
Thermostat 137
Time delay 42
Time delay fuses 145
Total lift 95
Trenching 125
Tuberculosis 25
Turbidity 55, 67, 68, 70
Typhoid fever 33

U

Ultra-violet light 35, 41, 42
Ultra-violet tubes 42
Underground iron deposits 52
Underground water 50
Underground water sources 26
Undissolved materials 47
United States Environmental Protection
    Agency 54
Units
  water conditioning 63
  water softening 63

V

Valve 16, 20
Velocity 38
Ventilation 116
Venturi 85
Venturi unit 111
Viruses 25, 43, 44, 46
Voltage 42, 141

W

Water
  demineralized 47
  distilled 47
  ground 50
  hard 47, 61
  pond 66
  safe 25
  soft 47
  surface 33
  untreated 40
Watering
  livestock 11
  poultry 11
Waterers 135
  livestock 137
  poultry 137
Water conditioning 51, 53
Water conditioning equipment 93
Water conditioning needs 57
Water conditioning unit 57
Water hammer 113
Water hardness 58
Water-level indicator 18
Water pasteurizers 39
Water sample 33
Water softeners 50, 61, 62, 65
Water storage 97
Water-supply control switches ... 105
Water supply yield 100
Water turbine 42
Water yield 16
Wearing ring 83
Weighted float 17
Wells
  bored 13, 14, 16, 19, 23, 26, 28
  drilled 13, 16, 22, 28
  driven 13, 14, 16, 19, 23
  dug 13, 14, 16, 19, 23, 26, 28, 49
  open dug 50
  jetted 13, 14, 16, 19, 23
  open 33
Well casing 26
Well drilling machines
  percussion type 13
  rotary type 13
Well size 87
Windmills 139, 140
Wire
  aluminum circuit 142
  copper circuit 142
  low-wattage heater 137
  service 146
  underground 142
  underground wiring 143
Wire size 141

Z

Zeolite 59, 60, 61, 62, 63
Zeolite bags 61
Zeolite bed 63
Zeolite minerals 63
Zeotone 62

Notes

Notes

Notes